T0155579

Advanced Forecasting with Python

With State-of-the-Art-Models Including LSTMs, Facebook's Prophet, and Amazon's DeepAR

Joos Korstanje

Apress®

Advanced Forecasting with Python: With State-of-the-Art-Models Including LSTMs, Facebook's Prophet, and Amazon's DeepAR

Joos Korstanje
Maisons Alfort, France

ISBN-13 (pbk): 978-1-4842-7149-0   ISBN-13 (electronic): 978-1-4842-7150-6
https://doi.org/10.1007/978-1-4842-7150-6

Managing Director, Apress Media LLC: Welmoed Spahr
Acquisitions Editor: Celestin Suresh John
Development Editor: Matthew Moodie
Coordinating Editor: Aditee Mirashi

Cover designed by eStudioCalamar

Cover image designed by Freepik (www.freepik.com)

Distributed to the book trade worldwide by Springer Science+Business Media New York, 1 New York Plaza, Suite 4600, New York, NY 10004-1562, USA. Phone 1-800-SPRINGER, fax (201) 348-4505, e-mail orders-ny@ springer-sbm.com, or visit www.springeronline.com. Apress Media, LLC is a California LLC and the sole member (owner) is Springer Science + Business Media Finance Inc (SSBM Finance Inc). SSBM Finance Inc is a **Delaware** corporation.

For information on translations, please e-mail booktranslations@springernature.com; for reprint, paperback, or audio rights, please e-mail bookpermissions@springernature.com.

Apress titles may be purchased in bulk for academic, corporate, or promotional use. eBook versions and licenses are also available for most titles. For more information, reference our Print and eBook Bulk Sales web page at http://www.apress.com/bulk-sales.

Any source code or other supplementary material referenced by the author in this book is available to readers on GitHub via the book's product page, located at www.apress.com/978-1-4842-7149-0. For more detailed information, please visit http://www.apress.com/source-code.

Printed on acid-free paper

This book is dedicated to my partner, Olivia,
for the help and support throughout the period of writing.

Table of Contents

About the Author xiii

About the Technical Reviewer xv

Introduction xvii

Part I: Machine Learning for Forecasting 1

Chapter 1: Models for Forecasting 3

Reading Guide for This Book 4

Machine Learning Landscape 4

    Univariate Time Series Models 4

    Supervised Machine Learning Models 9

    Other Distinctions in Machine Learning Models 17

Key Takeaways 18

Chapter 2: Model Evaluation for Forecasting 21

Evaluation with an Example Forecast 21

Model Quality Metrics 24

    Metric 1: MSE 25

    Metric 2: RMSE 26

    Metric 3: MAE 27

    Metric 4: MAPE 28

    Metric 5: R2 28

Model Evaluation Strategies 29

    Overfit and the Out-of-Sample Error 30

    Strategy 1: Train-Test Split 30

    Strategy 2: Train-Validation-Test Split 32

    Strategy 3: Cross-Validation for Forecasting 34

Backtesting 39

Which Strategy to Use for Safe Forecasts? 40

Final Considerations on Model Evaluation 41

Key Takeaways 42

Part II: Univariate Time Series Models 43

Chapter 3: The AR Model 45

Autocorrelation: The Past Influences the Present 46

    Compute Autocorrelation in Earthquake Counts 46

    Positive and Negative Autocorrelation 50

Stationarity and the ADF Test 51

Differencing a Time Series 52

Lags in Autocorrelation 55

    Partial Autocorrelation 57

    How Many Lags to Include? 58

AR Model Definition 59

Estimating the AR Using Yule-Walker Equations 60

    The Yule-Walker Method 60

    Train-Test Evaluation and Tuning 64

Key Takeaways 69

Chapter 4: The MA Model 71

The Model Definition 72

Fitting the MA Model 73

Stationarity 74

Choosing Between an AR and an MA Model 74

Application of the MA Model 75

Multistep Forecasting with Model Retraining 82

Grid Search to Find the Best MA Order 84

Key Takeaways 86

Chapter 5: The ARMA Model 89

The Idea Behind the ARMA Model 89

The Mathematical Definition of the ARMA Model 90

An Example: Predicting Sunspots Using ARMA 90

Fitting an ARMA(1,1) Model 94

More Model Evaluation KPIs 96

Automated Hyperparameter Tuning 99

Grid Search: Tuning for Predictive Performance 100

Key Takeaways 104

Chapter 6: The ARIMA Model 105

ARIMA Model Definition 106

Model Definition 106

ARIMA on the CO2 Example 107

Key Takeaways 113

Chapter 7: The SARIMA Model 115

Univariate Time Series Model Breakdown 115

The SARIMA Model Definition 116

Example: SARIMA on Walmart Sales 117

Key Takeaways 122

Part III: Multivariate Time Series Models 123

Chapter 8: The SARIMAX Model 125

Time Series Building Blocks 125

Model Definition 126

Supervised Models vs. SARIMAX 126

Example of SARIMAX on the Walmart Dataset 127

Key Takeaways 131

Chapter 9: The VAR Model **133**

The Model Definition 133

    Order: Only One Hyperparameter 134

    Stationarity 134

    Estimation of the VAR Coefficients 135

One Multivariate Model vs. Multiple Univariate Models 135

An Example: VAR for Forecasting Walmart Sales 136

Key Takeaways 139

Chapter 10: The VARMAX Model **141**

Model Definition 142

Multiple Time Series with Exogenous Variables 142

Key Takeaways 145

Part IV: Supervised Machine Learning Models **147**

Chapter 11: The Linear Regression **149**

The Idea Behind Linear Regression 150

Model Definition 150

Example: Linear Model to Forecast CO_2 Levels 151

Key Takeaways 157

Chapter 12: The Decision Tree Model **159**

Mathematics 160

    Splitting 160

    Pruning and Reducing Complexity 160

Example 161

Key Takeaways 168

Chapter 13: The kNN Model **169**

Intuitive Explanation 169

Mathematical Definition of Nearest Neighbors 169

    Combining k Neighbors into One Forecast 171

    Deciding on the Number of Neighbors k 171

Predicting Traffic Using kNN 172

Grid Search on kNN 175

Random Search: An Alternative to Grid Search 176

Key Takeaways... 177

Chapter 14: The Random Forest 179

Intuitive Idea Behind Random Forests 179

Random Forest Concept 1: Ensemble Learning 180

    Bagging Concept 1: Bootstrap 180

    Bagging Concept 2: Aggregation 181

Random Forest Concept 2: Variable Subsets 182

Predicting Sunspots Using a Random Forest... 182

Grid Search on the Two Main Hyperparameters of the Random Forest 184

Random Search CV Using Distributions 185

    Distribution for max_features.. 186

    Distribution for n_estimators.. 187

    Fitting the RandomizedSearchCV 188

Interpretation of Random Forests: Feature Importance 189

Key Takeaways... 191

Chapter 15: Gradient Boosting with XGBoost and LightGBM 193

Boosting: A Different Way of Ensemble Learning 193

The Gradient in Gradient Boosting 194

Gradient Boosting Algorithms 195

The Difference Between XGBoost and LightGBM... 195

Forecasting Traffic Volume with XGBoost.. 197

Forecasting Traffic Volume with LightGBM 199

Hyperparameter Tuning Using Bayesian Optimization 200

    The Theory of Bayesian Optimization 201

    Bayesian Optimization Using scikit-optimize 202

Conclusion 204

Key Takeaways... 205

Part V: Advanced Machine and Deep Learning Models 207

Chapter 16: Neural Networks 209

Fully Connected Neural Networks.. 209

Activation Functions.. 211

The Weights: Backpropagation.. 211

Optimizers.. 212

Learning Rate of the Optimizer 212

Hyperparameters at Play in Developing a NN 213

Introducing the Example Data.. 214

Specific Data Prep Needs for a NN 215

    Scaling and Standardization.. 215

    Principal Component Analysis (PCA)... 216

The Neural Network Using Keras 219

Conclusion 225

Key Takeaways... 226

Chapter 17: RNNs Using SimpleRNN and GRU 227

What Are RNNs: Architecture... 227

Inside the SimpleRNN Unit.. 228

The Example 229

Predicting a Sequence Rather Than a Value 230

Univariate Model Rather Than Multivariable 230

Preparing the Data 230

A Simple SimpleRNN.. 233

SimpleRNN with Hidden Layers 235

Simple GRU 237

GRU with Hidden Layers... 240

Key Takeaways... 242

Chapter 18: LSTM RNNs...**243**

    What Is LSTM 243

    The LSTM Cell 243

    Example 244

    LSTM with One Layer of 8 246

    LSTM with Three Layers of 64 248

    Conclusion 251

    Key Takeaways.. 251

Chapter 19: The Prophet Model **253**

    The Example 254

    The Prophet Data Format.. 254

    The Basic Prophet Model 255

    Adding Monthly Seasonality to Prophet.................................... 259

    Adding Holiday Data to Basic Prophet 260

    Adding an Extra Regressor to Prophet...................................... 263

    Tuning Hyperparameters Using Grid Search 266

    Key Takeaways.. 271

Chapter 20: The DeepAR Model **273**

    About DeepAR 273

    Model Training with DeepAR 274

    Predictions with DeepAR.. 276

    Probability Predictions with DeepAR.. 277

    Adding Extra Regressors to DeepAR 279

    Hyperparameters of the DeepAR.. 281

    Benchmark and Conclusion 283

    Key Takeaways.. 284

Chapter 21: Model Selection...**285**

Model Selection Based on Metrics.. 285

Model Structure and Inputs 286

One-Step Forecasts vs. Multistep Forecasts 287

Model Complexity vs. Gain.. 287

Model Complexity vs. Interpretability... 288

Model Stability and Variation 289

Conclusion 289

Key Takeaways... 290

Index...**291**

About the Author

 **Joos Korstanje** is a data scientist, with over five years of industry experience in developing machine learning tools, of which a large part is forecasting models. He currently works at Disneyland Paris where he develops machine learning for a variety of tools. His experience in writing and teaching has motivated him to write this book, *Advanced Forecasting with Python*.

About the Technical Reviewer

 **Michael Keith** is a data scientist working in the public health sector based in Salt Lake City, Utah. He is passionate about using data to improve health and educational outcomes and is a lead forecaster for the Utah Department of Health, leveraging Python to produce hundreds of forecasts every month. He earned a master's degree from Florida State University and has worked in data-related roles for several organizations, including Disney in Orlando. He has produced data science–themed videos for Apress, writes for *Towards Data Science*, performs consultations for Western Governors University, and lectures annually to graduate students at Florida State. In his free time, he enjoys road biking, hiking, and watching movies with his wife and beautiful 7-month-old daughter.

Introduction

Advanced Forecasting with Python covers all machine learning techniques relevant for forecasting problems, ranging from univariate and multivariate time series to supervised learning, to state-of-the-art deep forecasting models like LSTMs, Recurrent Neural Networks (RNNs), Facebook's open source Prophet model, and Amazon's DeepAR model.

Rather than focus on a specific set of models, this book presents an exhaustive overview of all techniques relevant to practitioners of forecasting. It begins by explaining the different categories of models that are relevant for forecasting in a high-level language. Next, it covers univariate and multivariate time series models followed by advanced machine learning and deep learning models, such as Recurrent Neural Networks, LSTMs, Facebook's Prophet, and Amazon's DeepAR. It concludes with reflections on model selection like benchmark scores vs. understandability of models vs. compute time and automated retraining and updating of models. Each of the models presented in this book is covered in depth, with an intuitive simple explanation of the model, a mathematical transcription of this idea, and Python code that applies the model to an example dataset.

This book is a great resource for those who want to add a competitive edge to their current forecasting skillset. The book is also adapted to those who start working on forecasting tasks and are looking for an exhaustive book that allows them to start with traditional models and gradually move into more and more advanced models.

You can follow along with the code using the GitHub repository that contains a Jupyter notebook per chapter. You are encouraged to use Jupyter notebooks for following along, but you can also run the code in any other Python environment of your choice.

PART I

Machine Learning for Forecasting

CHAPTER 1

Models for Forecasting

Forecasting, grossly translated as the task of predicting the future, has been present in human society for ages. Whether it is through fortune-tellers, weather forecasts, or algorithmic stock trading, man has always been interested in predicting what the future holds.

Yet forecasting the future is not easy. Consider fortune-tellers, stock market gurus, or weather forecasters: many try to predict the future, but few succeed. And for those who succeed, you will never know whether it was luck or skill.

In recent years, the computing power of computers has become much more commonly available than, say, 30 years ago. This has created a great boom in the use of Artificial Intelligence. Artificial Intelligence and especially machine learning can be used for a wide range of tasks, including robotics, self-driving cars, but also forecasting, that is, if you have a reasonable amount of data about the past that you can project into the future.

Throughout this book, you will learn the modern machine learning techniques that are relevant for forecasting. I will present a large number of machine learning models, together with an intuitive explanation of the model, its mathematics, and an applied use case.

The goal of this book is to give you a real insight into the application of those machine learning models. You will see worked examples applied to real datasets together with honest evaluations of the results: some successful, some less successful.

In this way, this book is different than many other resources, which often present perfectly fitting use cases on simulated data. To learn real-life machine learning and forecasting, it is important to know how models work, but it is even more important to know how to evaluate a model honestly and objectively. This pragmatical point of view will be the guideline throughout the chapters.

© Joos Korstanje 2021
J. Korstanje, *Advanced Forecasting with Python*, https://doi.org/10.1007/978-1-4842-7150-6_1

Reading Guide for This Book

Before going further into the different models throughout this book, I will first present a general overview of the machine learning landscape: many types and families of models exist. Each of them has its applications. Before starting, it is important to have an overview of the types of models that exist in machine learning and which of them are relevant for forecasting.

After this, I will cover several strategies and metrics for evaluating forecasting models. It is important to understand objective evaluation before practicing: you need to understand your goal before starting to practice.

The remaining chapters of the book will each cover a specific model with an intuitive explanation of the model, its mathematical definitions, and an application on a real dataset. You will start simple with common but simple methods and work your way up to the most recent and state-of-the-art methods on the market.

Machine Learning Landscape

Having the bigger picture of machine learning models before getting into detail will help you to understand how the different models compare to each other and will help you to keep the big picture throughout the book. You will first see univariate time series and supervised regression models: the main categories of forecasting models. After that, you will see a shorter description of machine learning techniques that are less relevant for forecasting.

Univariate Time Series Models

The first category of machine learning models that I want to talk about is time series models. Even though univariate time series have been around for a long time, they are still used. They also form an important basis for several state-of-the-art techniques. They are classical techniques that any forecaster should be familiar with.

Time series models are models that make a forecast of a variable by looking only at historical developments of the variable itself. This means that time series, as opposed to other model families, do not try to describe any "logical" relationships between variables. They do not try to explain the "why" of trends or seasonalities, but they simply put a mathematical formula on the past and try to project it to the future.

Time series modeling is sometimes criticized for this "lack of science." But time series models have gained an important place in forecasting due to their performances, and they could not be ignored.

A Quick Example of the Time Series Approach

Let's look at a super-simple, purely hypothetical example of forecasting the average price of a cup of coffee in an imaginary city called X. Imagine someone has made the effort of collecting the average price of coffee for 90 years in this town, with intervals of five years, and that this has yielded the data in Table 1-1.

Table 1-1. *A Hypothetical Example: The Price of a Cup of Coffee Over the Years*

Year	Average Price
1960	0.80
1965	1.00
1970	1.20
1975	1.40
1980	1.60
1985	1.80
1990	2.00
1995	2.20
2000	2.40
2005	2.60
2010	2.80
2015	3.00
2020	3.20

This fictitious data clearly shows an increase of 20 cents in the price every five years. This is a **linear increasing trend**: linear because it increases with the same amount every year and increasing because it becomes more rather than less.

Let's get this data into Python to see how to plot this linear increasing trend using Listing 1-1. The source code for this book is available on GitHub via the book's product page, located at www.apress.com/978-1-4842-7149-0. Please note that the library imports are done once per chapter.

Listing 1-1. Getting the coffee example into Python and plotting the trend

```python
import pandas as pd
import matplotlib.pyplot as plt

years = [1965, 1970, 1975, 1980, 1985, 1990, 1995, 2000, 2005, 2010, 2015, 2020]
prices = [1.00, 1.20, 1.40, 1.60, 1.80, 2.00, 2.20, 2.40, 2.60, 2.80, 3.00, 3.20]

data = pd.DataFrame({
    'year' : years,
    'prices': prices
})
ax = data.plot.line(x='year')
ax.set_title('Coffee Price Over Time', fontsize=16)
plt.show()
```

You will obtain the graph displayed in Figure 1-1.

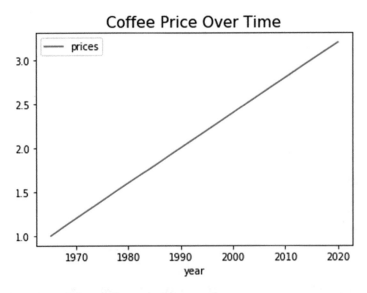

Figure 1-1. *The plot of the coffee price example*

To make predictions for the price of coffee in this hypothetical town, you could just put your ruler next to the graph and continue the upward line: the prediction for this variable does not need any **explanatory variables** other than its past values. The historical data of this example allows you to forecast the future. This is a determining characteristic of **time series models**.

Now let's see a comparable example but with the prices of hot chocolate rather than the prices of a cup of coffee and quarterly data rather than data every five years (Table 1-2).

Table 1-2. *Hot Chocolate Prices Over the Years*

Period	Average Price
Spring 2018	2.80
Summer 2018	2.60
Autumn 2018	3.00
Winter 2018	3.20
Spring 2019	2.80
Summer 2019	2.60
Autumn 2019	3.00
Winter 2019	3.20
Spring 2020	2.80
Summer 2020	2.60
Autumn 2020	3.00
Winter 2020	3.20

Do you see the trend? In the case of hot chocolate, you do not have a year-over-year increase in price, but you do detect **seasonality**: in the example, hot chocolate prices follow the temperatures of the seasons. Let's get this data into Python to see how to plot this seasonal trend (use Listing 1-2 to obtain the graph in Figure 1-2).

Listing 1-2. Getting the hot chocolate example into Python and plotting the trend

```
seasons = ["Spring 2018", "Summer 2018", "Autumn 2018", "Winter 2018",
           "Spring 2019", "Summer 2019", "Autumn 2019", "Winter 2019",
           "Spring 2020", "Summer 2020", "Autumn 2020", "Winter 2020"]
prices = [2.80, 2.60, 3.00, 3.20,
          2.80, 2.60, 3.00, 3.20,
          2.80, 2.60, 3.00, 3.20]

data = pd.DataFrame({
    'season': seasons,
    'price': prices
})

ax = data.plot.line(x='season')
ax.set_title('Hot Chocolate Price Over Time', fontsize=16)
ax.set_xticklabels(ax.get_xticklabels(), rotation=45, ha='right')
plt.show()
```

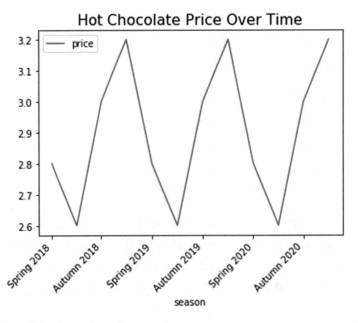

Figure 1-2. *Plot of the hot chocolate prices*

As in the previous example, you can predict the future prices of hot chocolate easily using the past data on hot chocolate prices: the prices depend only on the season and are not influenced by any explanatory variables.

Note   Univariate time series models make predictions based on trends and seasonality observed in their own past and do not use explanatory variables other than the **target variable**: **the variable that you want to forecast**.

You can imagine numerous types of combinations of those two processes, for example, have both a quarterly seasonality and a linear increasing trend and so on. There are many types of processes that can be forecasted by modeling the historical values of the target variable. In Chapters 3–7, you will see numerous univariate time series models for forecasting.

Supervised Machine Learning Models

Now that you are familiar with the idea of using the past of one variable, you are going to discover a different approach to making models. You have just seen univariate time series models, which are models that use only the past of a variable itself to predict its future.

Sometimes, this approach is not logical: processes do not always follow trends and seasonality. Some predictions that you would want to make may be dependent on other, independent sources of information: **explanatory variables**.

In those cases, you can use a family of methods called **supervised machine learning** that allows you to model relationships between explanatory variables and a target variable.

A Quick Example of the Supervised Machine Learning Approach

To understand this case, you have the fictitious data in Table 1-3: a new example that contains the sales amount of a company per quarter, with three years of historical data.

Table 1-3. *Quarterly Sales*

Period	Quarterly Sales
Q1 2018	48,000
Q2 2018	20,000
Q3 2018	35,000
Q4 2018	32,0000
Q1 2019	16,000
Q2 2019	58,000
Q3 2019	40,000
Q4 2019	30,000
Q1 2020	32,000
Q2 2020	31,000
Q3 2020	63,000
Q4 2020	57,000

To get this data into Python, you can use the following code (Listing 1-3).

Listing 1-3. Getting the quarterly sales example into Python and plotting the trend

```
quarters = ["Q1 2018", "Q2 2018", "Q3 2018", "Q4 2018",
            "Q1 2019", "Q2 2019", "Q3 2019", "Q4 2019",
            "Q1 2020", "Q2 2020", "Q3 2020", "Q4 2020"]

sales = [48, 20, 42, 32,
         16, 58, 40, 30,
         32, 31, 53, 40]

data = pd.DataFrame({
    'quarter': quarters,
    'sales': sales
})
```

```
ax = data.plot.line(x='quarter')
ax.set_title('Sales Per Quarter', fontsize=16)
ax.set_xticklabels(ax.get_xticklabels(), rotation=45, ha='right')
plt.show()
```

The graph that you obtain is a line graph that shows the sales over time (Figure 1-3).

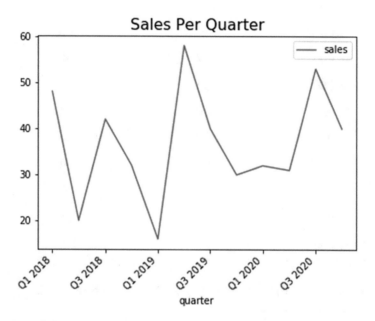

Figure 1-3. *Plot of the quarterly sales*

What you can see in this graph does not resemble the previous examples: there is no clear linear trend (neither increasing nor decreasing), and there is no clear quarterly seasonality either.

But as the data is about sales, you could imagine many factors that influence the sales that you'll realize. Let's look for explanatory variables that could help in explaining sales. In Table 1-4, the data have been updated with two explanatory variables: discount and advertising budget. Both are potential variables that could influence sales numbers.

Table 1-4. *Quarterly Sales, Discount, and Advertising Budget*

Period	Quarterly Sales	Avg. Discount	Advertising Budget
Q1 2018	48,000	4%	500
Q2 2018	20,000	2%	150
Q3 2018	35,000	3%	400
Q4 2018	32,0000	3%	300
Q1 2019	16,000	2%	100
Q2 2019	58,000	6%	500
Q3 2019	40,000	4%	380
Q4 2019	30,000	3%	280
Q1 2020	32,000	3%	290
Q2 2020	31,000	3%	315
Q3 2020	63,000	6%	625
Q4 2020	57,000	6%	585

Let's have a look at whether it would be possible to use those variables for a prediction of sales using Listing 1-4.

Listing 1-4. Getting the quarterly sales example into Python and plotting the trend

```python
quarters = ["Q1 2018", "Q2 2018", "Q3 2018", "Q4 2018",
            "Q1 2019", "Q2 2019", "Q3 2019", "Q4 2019",
            "Q1 2020", "Q2 2020", "Q3 2020", "Q4 2020"]

sales = [48, 20, 42, 32,
         16, 58, 40, 30,
         32, 31, 53, 40]

discounts = [4,2,3,
             3,2,6,
             4,3,3,
             3,6,6]
```

```
advertising = [500,150,400,
               300,100,500,
               380,280,290,
               315,625,585]
data = pd.DataFrame({
    'quarter': quarters,
    'sales': sales,
    'discount': discounts,
    'advertising': advertising
})

ax = data.plot.line(x='quarter')
ax.set_title('Sales Per Quarter', fontsize=16)
ax.set_xticklabels(ax.get_xticklabels(), rotation=45, ha='right')
plt.show()
```

This gives you the graph that is displayed in Figure 1-4: a graph displaying the development of the three variables over time.

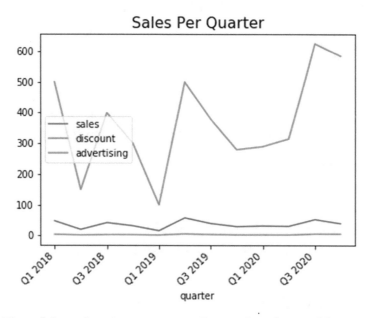

Figure 1-4. *Plot of the sales per quarter with correlated variables*

At this point, visually, you'd probably say that there is not a very important relationship between the three variables. But let's have a more zoomed-in look at the same graph (Listing 1-5).

Listing 1-5. Zooming in on the correlated variables of the quarterly sales example

```python
quarters = ["Q1 2018", "Q2 2018", "Q3 2018", "Q4 2018",
            "Q1 2019", "Q2 2019", "Q3 2019", "Q4 2019",
            "Q1 2020", "Q2 2020", "Q3 2020", "Q4 2020"]

sales = [48, 20, 42, 32,
         16, 58, 40, 30,
         32, 31, 53, 40]

discounts = [4,2,3,
             3,2,6,
             4,3,3,
             3,6,6]

discounts_scale_adjusted = [x * 10 for x in discounts]

advertising = [500,150,400,
               300,100,500,
               380,280,290,
               315,625,585]

advertising_scale_adjusted = [x / 10 for x in advertising]

data = pd.DataFrame({
    'quarter': quarters,
    'sales': sales,
    'discount': discounts_scale_adjusted,
    'advertising': advertising_scale_adjusted
})

ax = data.plot.line(x='quarter')
ax.set_title('Sales Per Quarter', fontsize=16)
ax.set_xticklabels(ax.get_xticklabels(), rotation=45, ha='right')
plt.show()
```

This gives the graph displayed in Figure 1-5: you can suddenly observe a very clear relationship between the three variables! The relationship was already there in the previous graph (Figure 1-4), but it was just not visually obvious due to the difference in scale of the curves.

Figure 1-5. *Zoomed view of the correlated variables of the quarterly sales example*

Imagine you observe a correlation as strong as in Figure 1-5. If you had to do this sales forecast for next month, you could simply ask your colleagues what the average discount is going to be next month and what next month's advertising budget is, and you would be able to come up with a reasonable guess of the future sales.

This type of relationships is what you are generally looking at when doing supervised machine learning. Intelligent use of those relations is the fundamental idea behind the different techniques that you will see throughout this book.

Correlation Coefficient

The visual way to detect correlation is great. Yet there is a more exact way to investigate relationships between variables: the correlation coefficient. The **correlation coefficient** is a very important measure in statistics and machine learning as it determines how much two variables are correlated.

The correlation coefficient between two variables x and y can be computed as follows:

A **correlation matrix** is a matrix that contains the correlations between each pair of variables in a dataset. Use Listing 1-6 to obtain a correlation matrix.

Listing 1-6. Getting the quarterly sales example into Python and plotting the trend

```
data.corr()
```

It will give you the correlations between each pair of variables in the dataset as shown in Figure 1-6.

	sales	discount	advertising
sales	1.000000	0.848135	0.902568
discount	0.848135	1.000000	0.920958
advertising	0.902568	0.920958	1.000000

Figure 1-6. *Correlation table of the quarterly sales example*

A correlation coefficient is always **between -1 and 1**. A positive value for the correlation coefficient means that two variables are positively correlated: if one is higher, then the other is generally also higher. If the correlation coefficient is negative, there is a negative correlation: if one value is higher, then the other is generally lower. This is the **direction of the correlation**.

There is also a notion of the **strength of the correlation**. A correlation that is close to 1 or close to -1 is strong. A correlation coefficient that is close to 0 is a weak correlation. Strong correlations are generally more interesting, as an explanatory variable that strongly correlated to your variable can be used for forecasting it.

In the variables in the example, you see a strong positive correlation between sales and discount (0.848) and a strong positive correlation between sales and advertising (0.90). As the correlations are strong, they could be useful in predicting future sales.

You can also observe a strong correlation between the two explanatory variables discount and advertising (0.92). This is important to notice, because if two explanatory variables are very correlated, it may be of little added value to use them both: they contain almost the same "information," just measured on a different scale.

Later, you'll discover numerous mathematical models that will allow you to make models of the relationships between a target variable and explanatory variables. This will help you choose which correlated variables to use in a predictive model.

Other Distinctions in Machine Learning Models

To complete the big picture overview of the machine learning landscape, there are a few more groups that need to be mentioned.

Supervised vs. Unsupervised Models

You have just seen what supervised models are all about. But there also unsupervised models. Unsupervised models differ from all approaches before, as there is no target variable in unsupervised models.

Unsupervised models are great for making segmentations. A classic example is regrouping customers of a store, based on the similarity of those customers. There are many great use cases for such segmentations, but the approach is not generally useful for forecasting.

Classification vs. Regression Models

Inside the group of supervised models, there is an important split between **classification** and **regression**. Regression is supervised modeling in which the target variable is **numeric**. In the examples that you have seen in the previous sections, there were only numeric target variables (e.g., amount of sales). They would therefore be considered regressions.

In classification, the target variable is **categorical**. An example of classification is predicting whether a specific customer will buy a product (yes or no), based on their customer behavior.

There are many important use cases for classification, but when talking about forecasting the future, it is generally numerical problems and therefore regression models. Think about the weather forecast: rather than trying to predict either hot or cold weather (which are categories of weather), you can try to predict the temperature (numeric). And instead of predicting rainy vs. dry weather (categories), you would predict the percentage of rain.

Univariate vs. Multivariate Models

A last important split between different approaches in machine learning is the split between **univariate models** and **multivariate models**.

In the examples you have seen until now, there was always one target variable (sales). But in some cases, you want to predict multiple related variables at the same time. For example, on social media, you may want to forecast how many likes you will receive, but also how many comments.

The first possibility for treating this is to build two models: one model for forecasting likes and another one for forecasting the number of comments. But some models allow benefiting from a correlation between target variables. These are called multivariate models, and they use the correlation between target variables in such a way that the forecast accuracy improves from forecasting the two at the same time.

You will see multivariate models in Chapters 8 and 9. Multivariate models can be great scientific descriptions of reality, but caution needs to be paid to their predictive performance. Multiple models are sometimes more performant than one model that makes multiple predictions. Data-driven model evaluation strategies will help you to make the best choice. This will be the scope of the next chapter.

Key Takeaways

1. Univariate time series models use historical data of a target variable to make predictions.

2. Seasonality and trend are important effects that can be used in time series models.

3. Supervised machine learning uses correlations between variables to forecast a target variable.

4. Supervised machine learning can be split into classification and regression. In classification, the target variable is categorical; in regression, the target variable is numeric. Regression is most relevant for forecasting.

5. The correlation coefficient is a KPI of the relationship between two variables. If the value is close to 1 or -1, the correlation is strong; if it is close to 0, it is weak. A strong correlation between an explanatory variable and the target variable is useful in supervised machine learning.

6. Univariate models predict one variable. Multivariate models predict multiple variables. Most forecasting models are univariate, but some multivariate models exist.

## CHAPTER 2

Model Evaluation for Forecasting

When developing machine learning models, you generally benchmark multiple models during the build phase. Then you estimate the performances of those models and select the model which you consider most likely to perform well. You need objective measures of performance to decide which forecast to retain as your actual forecast.

In this chapter, you'll discover numerous tools for model evaluation. You are going to see different strategies for evaluating machine learning models in general and specific adaptations and considerations to take into account for forecasting. You are also going to see different metrics for scoring model performances.

Evaluation with an Example Forecast

Let's look at a purely hypothetical example with stock prices per month of the year 2020 and the forecast that someone has made for this (Table 2-1). Assume that this forecast has been made in December 2019 for the complete year and that it has not been updated since.

© Joos Korstanje 2021
J. Korstanje, *Advanced Forecasting with Python*, https://doi.org/10.1007/978-1-4842-7150-6_2

Table 2-1. *Stock Price Data*

Period	Stock Price	Forecasted
January	35	30
February	35	31
March	10	30
April	5	10
May	8	12
June	10	17
July	15	18
August	20	27
September	23	29
October	21	24
November	22	23
December	25	22

You can already see that there is quite some difference between the actual values and the forecasted values. But that happens. Let's start with getting the data into Python and plotting the two lines using Listing 2-1.

Listing 2-1. Getting the stock data example into Python

```
import pandas as pd
import matplotlib.pyplot as plt

period = ['January', 'February', 'March',
        'April', 'May', 'June',
        'July', 'August', 'September',
        'October', 'November', 'December']

actual = [35, 35, 10,
        5, 8, 10,
        15, 20, 23,
        21, 22, 25]
```

```
forecast = [30, 31, 30,
            10, 12, 17,
            18, 27, 29,
            24, 23, 22]

data = pd.DataFrame({
    'period': period,
    'actual': actual,
    'forecast': forecast
})

ax = data.plot.line(x='period')
ax.set_title('Forecast vs Actual', fontsize=16)
ax.set_xticklabels(ax.get_xticklabels(), rotation=45, ha='right')
plt.show()
```

This should give you the graph in Figure 2-1, which displays the actual values against the forecasted values over time.

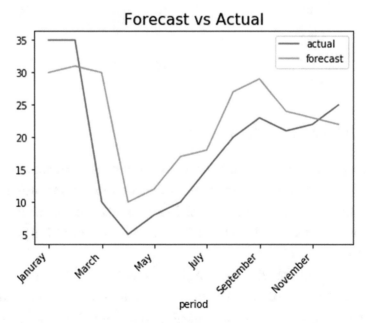

Figure 2-1. *Stock prices vs. forecasted stock prices over time*

Now the next simple step is to compute the differences between each forecasted value and each actual value, as shown in Table 2-2.

Table 2-2. *Adding the Errors of Each Forecasted Value to the Data*

Period	Stock Price	Forecasted	Error
January	35	30	-5
February	35	31	-4
March	10	30	20
April	5	10	5
May	8	12	4
June	10	17	7
July	15	18	3
August	20	27	7
September	23	29	6
October	21	24	3
November	22	23	1
December	25	22	-3

Model Quality Metrics

This month-by-month error is useful information for model improvement. Yet although it gives a first impression of the quality of the model, there are some problems with this way of model evaluation. Firstly, this level of detail would be too much to do a model comparison: you cannot look at long lists of errors for each model, but you ideally want one or a few KPIs per model.

Secondly, you should note that it is not simply possible to take the average of this Error column and consider this as your error metric. Since there are positive and negative error measures, they would average each other out. This would be a strong underestimation of your error.

To solve this problem, you need to standardize the errors by taking the absolute values or by squaring the errors. This will make sure that a negative and a positive error do not cancel each other out. You will now see five common metrics that do exactly this.

Metric 1: MSE

The Mean Squared Error (MSE) is one of the most used metrics in machine learning. It is computed as the average of the squared errors. To compute the MSE, you take the errors per row of data, square those errors, and then take the average of them.

$$MSE = \frac{1}{n}\Sigma\left(y_i - \hat{y}_i\right)^2$$

You can compute it in Python using the scikit-learn library, as is done in Listing 2-2.

Listing 2-2. Computing the Mean Squared Error in Python

```
from sklearn.metrics import mean_squared_error
print(mean_squared_error(data['actual'], data['forecast']))
```

For the current example, this gives a MSE of **53.7**.

The logic behind this error metric is multifold. First, you can understand the reason that the squared errors are used rather than the original errors, as it would be impossible to sum the original errors. Since there are positive and negative values in the original errors, they would cancel each other out. Imagine a case with one large negative error and one large positive error: the sum of the two errors might be close to zero, which is clearly wrong. The square of a value is always positive, which is why this is one possibility to counter this.

A second part of the formula that you can understand is that it functions as an average. You take the sum of all values and divide by the number of observations. In the MSE, you take the average of the squared errors.

The MSE error metric is great for comparing different models on the same dataset. The scale of the MSE will be the same for each model applied to the same dataset. However, the scale of the metric is not very intuitive, which makes it difficult to interpret outside of benchmarking multiple models.

The MSE is an error metric, so it should be interpreted as follows: the smaller the error, the better the model. It is not possible to convert this into an accuracy measure, because of the lack of a fixed scale of the metric. There are no upper bounds to the error, so it should really be used for comparison only.

Metric 2: RMSE

The RMSE, or Root Mean Squared Error, is the square root of the Mean Squared Error. As you can understand, taking the square root of the MSE does not make a difference when you want to use the error metrics for classing performances in order.

$$RMSE = \sqrt{MSE}$$

Yet there is an advantage in using the RMSE rather than the MSE. The reason for taking the square root of the MSE is that the scale of the RMSE is the same as the scale of the original variable. In the MSE formula, you take the average of squared errors. This makes the value difficult to interpret. Using the square root will get the scale of the error metric back to the scale of your actual values.

If you compute the RMSE using Listing 2-3, you will see that the RMSE is **7.3**. This is much more in line with the actual stock prices that you are working with. This can have an advantage for explanation and communication purposes.

Listing 2-3. Computing the Root Mean Squared Error in Python

```
from sklearn.metrics import mean_squared_error
from math import sqrt
print(sqrt(mean_squared_error(data['actual'], data['forecast'])))
```

As the RMSE is an error measure, a lower RMSE indicates a better model.

Although the RMSE is more intuitively understandable, its scale is still dependent on the actual values. This makes it impossible to compare the RMSE values of different datasets with one another, just like the MSE.

Metric 3: MAE

The Mean Absolute Error (MAE) is calculated by taking the absolute differences between the predicted and actual values per row. The average of those absolute errors is the Mean Absolute Error.

$$MAE = \frac{1}{n}\sum |y_i - \hat{y}_i|$$

The MAE takes the absolute values of the errors before averaging. Taking the average of absolute errors is a way to make sure that summing the errors will not make them cancel each other out.

You have seen the MSE using the square of the errors to avoid this, and the MAE is an alternative to this. The MAE has a more intuitive formula: it is the error metric that most people intuitively come up with. Yet the RMSE is generally favored over the MAE.

Since the RMSE uses squares rather than absolute values, the RMSE is easier to use in mathematical computations that demand to take derivatives. The derivative of squared errors is much easier to compute than the derivative of absolute errors. Since the derivative is a much used function in optimization and minimization, this is an important criterion.

The interpretation of the MAE is comparable to the interpretation of the RMSE. They both yield scores that are in the same range of values as the actual values. There will always be a difference in the MAE and the MSE. When using the squared errors, if one of the individual errors is very high, this value may weigh stronger in the total evaluation. Yet there is not a definite way to judge whether one of the error measures is better or worse than the other.

You can compute the Mean Absolute Error in Python using the code in Listing 2-4. You should obtain an MAE of **5.67**.

Listing 2-4. Computing the Mean Absolute Error in Python

```
from sklearn.metrics import mean_absolute_error
print(mean_absolute_error(data['actual'], data['forecast']))
```

Metric 4: MAPE

The MAPE, short for Mean Absolute Percent Error, is computed by taking the error for each prediction, divided by the actual value. This is done to obtain the errors relative to the actual values. This will make for an error measure as a percentage, and therefore it is standardized.

As you've understood from the previous error measures, they were not standardized on a scale between zero and one. Yet this standardization is very useful. This makes for very easy communication of the performance results.

To compute the MAPE, you take the absolute values of those percentages per row and compute their average. For the example in Listing 2-5, you'll obtain **0.46**.

$$MAPE = \frac{1}{n}\sum \left| \frac{y_i - \hat{y}_i}{y_i} \right|$$

The MAPE measures a percentage error. It is an error measure, so lower values for the MAPE are better. Yet you can easily convert the MAPE to a goodness of fit measure by computing 1 – MAPE. In many cases, it is easier to communicate performance in terms of a positive result rather than a negative one.

Although the MAPE function is intuitive, it has a serious drawback: when the actual value is 0, the formula will divide the error by the actual value. This leads to a division by zero, which is mathematically impossible. This makes it problematic to use.

Listing 2-5. Computing the Mean Absolute Percent Error in Python

```
from sklearn.metrics import mean_absolute_percentage_error
print(mean_absolute_percentage_error(data['actual'], data['forecast']))
```

Metric 5: R2

The R2 (R squared) metric is a metric that is very close to the 1 – MAPE metric. It is a performance metric rather than an error metric, which makes it great for communication.

The R2 is a value that tends to be between 0 and 1, with 0 being bad and 1 being perfect. It can therefore be easily used as a percentage by multiplying it by 100. The only case where the R2 can be negative is if your forecast is more than 100% wrong.

$$R^2 = 1 - \frac{\sum(y_i - \hat{y}_i)^2}{\sum(y_i - \bar{y}_i)^2}$$

The formula does an interesting computation. It computes a ratio between the sum of squared errors and the sum of deviations between the forecast and the average. This comes down to a percentage of increase of your model over using the average as a model. If your model is as bad a prediction as using the average, then the R2 will be zero. As the average is often used as a benchmark model, this is a very practical performance metric.

In Python, you can compute the R2 by using the code listed in Listing 2-6.

Listing 2-6. Computing the R2 in Python

```
from sklearn.metrics import r2_score
r2_score(data['actual'], data['forecast'])
```

This gives you a value (rounded) of **0.4**. As the R squared is measured on a scale between 0 and 1, you could translate this as 40% better performance than the average. Although this exact interpretation will not be very clear for your business partners and managers, the use of percentages to track model performance will be very convincing in many cases.

Model Evaluation Strategies

Now that you have seen five important metrics, let's look at how to set up tests for model comparison. When you're doing advanced forecasting, you will often be working with a lot of models at the same time. There are a lot of models that you can use, and the remainder of this book is going to present them.

But when working with all those different models, you generally need to make a final forecast: *how to decide which model is most accurate?*

In theory, you could of course use multiple models to predict the (short) future and then wait and see which one works best. This does happen in practice, and this is a great way to make sure you deliver quality forecasts. However, you can do more. Rather than waiting, it is much more interesting to try and use past data to estimate errors.

In practice, the question is: *how to decide which model is most accurate, without waiting for the future to confirm your model?*

Overfit and the Out-of-Sample Error

When doing forecasting, you are building models on historical data and projecting a forecast into the future. When doing this, it is important to avoid certain biases. A very common bias is to fit a model on historical data, compute errors on this historical data, and, if the error is good, use it for a forecast.

Yet, this will not work due to overfitting. Overfitting a model means that your model has learned the past data too specifically. With the advanced methods that are available on the market, models could get to fit almost any historical trend to 100%. However, this is not at all a guarantee for performance in out-of-sample prediction.

When models overfit, they obtain very high scores on historical data and poor performances on future data. Rather than learning general, true trends, an overfitted model remembers a lot of "useless" and "noisy" variations in the past and will project this noise into the future.

Machine learning models are powerful learning models: they will learn anything that you give them. But when they overfit, they learn too much. This happens often, and many strategies exist to avoid it.

Strategy 1: Train-Test Split

A first split that is often deployed is the train-test split. When applying a train-test split, you split the rows of data in two. You would generally keep 20% or 30% of data in a test set.

You can then proceed to fit models on the rest of the data: the training data. You can apply many models to the train data and predict on the test data. You compare the performance of the machine learning models on the test set, and you will notice that the models with good performance on the train data do not necessarily have a good performance on the test data.

A model with a higher score on the training data and a lower score on the test data is a model with an overfit. In this case, a model with a slightly lower error on the training data may well outperform the first model on the test set.

The test performances are those that matter, as they best replicate the future case: they try to predict on data that is "unknown" to the model – this replicates the situation of your forecast in the future.

Now, a question that you could ask is how to choose the test set. In most machine learning problems, you do a random selection of data for the test set. But forecasting is quite particular in this case. As you generally have an order in the data, it makes more sense to keep the test set as the 20% last observations: this will replicate the future situation of application of the forecast.

As an example of applying the train-test split, let's make a very simple forecasting model: the mean model. It consists of taking the average on the train data and using that as a forecast. Of course, it will not be very performant, but it is often used as a "minimum" benchmark in forecasting models. Any model that is worse than this is not worth being considered at all.

So let's see an example with the stock price data from before using Listing 2-7.

Listing 2-7. Train-test split in Python

```
from sklearn.model_selection import train_test_split
y = data['actual']
train, test = train_test_split(y, test_size=0.3, shuffle=False)
forecast = train.mean() # forecast is 17.25
train = pd.DataFrame(train)
train['forecast'] = forecast
train_error = mean_squared_error(train['actual'], train['forecast'])

test = pd.DataFrame(test)
test['forecast'] = forecast
test_error = mean_squared_error(test['actual'], test['forecast'])
print(train_error, test_error)
```

This will give a train error of 122.9375 and a test error of 32.4375. Attention here: With one year of data, a train-test split will cause a problem. The training data will not have a full year of data, which makes the model to not fully learn any seasonality. In addition, you should strive for having at least three observations per period (in this case, three years of training data).

It is generally a good idea to take seasonality into account when selecting the testing period. In non-forecasting situations, you will often see test sets of 20% or 30% of the data. In forecasting, I advise using a full seasonal period. If you have yearly seasonality, you should use a full year as a test period, to avoid having, for example, a model that works very well in summer, but badly in winter (potentially due to different dynamics in what you are forecasting in different seasons).

Strategy 2: Train-Validation-Test Split

When doing a **model comparison**, you benchmark the performances of many models. As said before, you can avoid overfitting by training the models on the train set and testing the models on the test set.

Adding to this approach, you can add an extra split: the **validation split**. When using the train-validation-test split, you will train models on the training data, then you benchmark the models on the validation data, and this will be the basis for your model selection. Then finally, you use your selected model to compute an error on the test set: this should confirm the estimated error of your model.

In case you have an error on the test set that is significantly worse than the error on the validation set, this will alert you that the model is not as good as you expected: the validation error is underestimated and should be reinvestigated.

When using the train-validation-test split, there is a serious amount of data dedicated fully to model comparison and testing. Therefore, this approach should be avoided when working with little data.

As said before, it is important to consider which periods you leave out for your validation and test data. If, for example, you have five years of monthly data and you decide to use three years for training, one year for validation, and one year for testing, you are missing out on the two most recent years of data: not a good idea.

In this case, you could select the best model (that is the **model type** and its optimal **hyperparameters**); and, once the model is selected and benchmarked, you retrain the optimal model including the most recent data.

Listing 2-8 shows a very basic example in which two models are compared: the mean and the median. Of course, those models should not be expected to be performant: the goal here is to show how to use the validation data for model comparison.

Listing 2-8. Train-validation-test split in Python

```python
# Splitting into 70% train, 15% validation and 15% test
train, test = train_test_split(data['actual'], test_size = 0.3,
shuffle = False, random_state=12345)
val, test = train_test_split(test, test_size = 0.5, shuffle = False,
random_state=12345)
```

```
# Fit (estimate) the two models on the train data
forecast_mean = train.mean() # 17.25
forecast_median = train.median() # 12.5

# Compute MSE on validation data for both models
val = pd.DataFrame(val)

val['forecast_mean'] = forecast_mean
val['forecast_median'] = forecast_median

mean_val_mse = mean_squared_error(val['actual'], val['forecast_mean'])
median_val_mse = mean_squared_error(val['actual'], val['forecast_median'])

# You observe the following validation mse: mean mse: 23.56,
median mse: 91.25
print(mean_val_mse, median_val_mse)

# The best performance is the mean model, so verify its error on test data
test = pd.DataFrame(test)
test['forecast_mean'] = forecast_mean

mean_test_mse = mean_squared_error(test['actual'], test['forecast_mean'])

# You observe a test mse of 41.3125, almost double the validation mse
print(mean_test_mse)
```

If you follow the example, you observe a validation MSE for the mean model of **23.56** and **91.25** for the median model. This would be a reason to retain the mean model against the median model. As a final evaluation, you verify whether the validation error is not influenced by a selection of the validation set that is (randomly) favorable to the mean model.

This is done by taking a final error measure on the test set. As both the validation and the test set were unseen data for the mean model, their errors should be close. However, you observe a test MSE of **41.3125**, almost double the validation error. The fact that the test error is far off from the validation error tells you that there is a bias in your error estimate. In this particular case, the bias is rather easy to find: there are too few data points in the validation set (only two data points), which makes the error estimate not reliable.

The train-validation-test set would have given you an early warning on the error metrics of your model. It would therefore have prevented you from relying on this forecast for future estimates. Although an evaluation strategy cannot improve a model, it can definitely improve your choice of forecasting models! It therefore has an indirect impact on your model's accuracy.

Strategy 3: Cross-Validation for Forecasting

A problem with the train-test set may occur when you have very little data. In some cases, you cannot "afford" to keep 20% of the data apart from the model. In this case, cross-validation is a great solution.

K-Fold Cross-Validation

The most common type of cross-validation is K-fold cross-validation. K-fold cross-validation is like an addition to the train-test split. In the train-test split, you compute the error of each model one time: on the test data. In K-fold cross-validation, you fit the same model k times, and you evaluate the error each time. You then obtain k error measures, of which you take the average. This is your cross-validation error.

To do this, a K-fold cross-validation makes a number of different train-test splits. If you choose a k of 10, you will split the data into ten. Then each of the ten parts will be serving as test data one time. This means that the remaining nine parts are used as training data.

As values for k, you are generally looking between 3 and 10. If you go much higher, the test data becomes small, which can lead to biased error estimates. If you go too low, you have very little folds, and you lose the added value of cross-validation. Note that when k is 1, you are simply applying a train-test split.

As a hypothetical example, imagine a case with 100 data points and a fivefold cross-validation. You will make five different train-test datasets, as you can see in the schematic illustration in Figure 2-2.

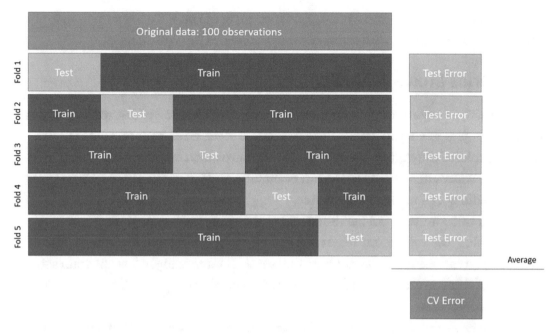

Figure 2-2. *K-fold cross-validation in Python*

On those five train-test splits, you train the model on the observations that have been selected as training data, and you compute the error on the observations that have been selected as the test data. The cross-validation error is then the average of the five test errors (Listing 2-9).

Listing 2-9. K-fold cross-validation in Python

```python
import numpy as np
from sklearn.model_selection import KFold

kf = KFold(n_splits=5)

errors = []
for train_index, test_index in kf.split(data):
    train = data.iloc[train_index,:]
    test = data.iloc[test_index,:]

    pred = train['actual'].mean()
    test['forecast'] = pred
    error = mean_squared_error(test['actual'], test['forecast'])
    errors.append(error)

print(np.mean(errors))
```

This example will give you a cross-validation error of **106.1**. This is a MSE, so it should only be compared with errors of other models on the same dataset.

Time Series Cross-Validation

The fact that you have multiple estimates of the error will improve the error estimate. As you can imagine, when you have only one estimate of the error, your test data may be very favorable: all difficult-to-predict events may have fallen in the training data! Cross-validation reduces this risk, by having an evaluation applied to all the data. This makes the error estimate generally more reliable.

Yet, attention must be paid to the specific case of forecasting. As you can see in the image, K-fold cross-validation creates test splits equally throughout the data. In the example, you can see many cases where the test set is temporally before the train set. This means that you are measuring the error of forecasting the past rather than the future using these methods.

For the category of supervised models, in which you use relations between your target variable and a set of explanatory variables, this is not necessarily a problem: the relationships between variables can be assumed to be the same in the past and the present.

However, for the category of time series models, this is a serious problem. Time series models are generally based on making forecasts based on trends and/or seasonality: they use the past of the target variable to forecast the future.

A first problem you'll encounter is that many time series models will not work with missing data: if a month is missing in the middle, the methods can simply not be estimated.

A second problem is that when the models can be estimated, they are often not realistically accurate: estimating a period in between data points is much easier than estimating a period that is totally in the future.

A solution that can be used is called **the time series split** (Figure 2-3). In this approach, you take only data that is before the test period for model training.

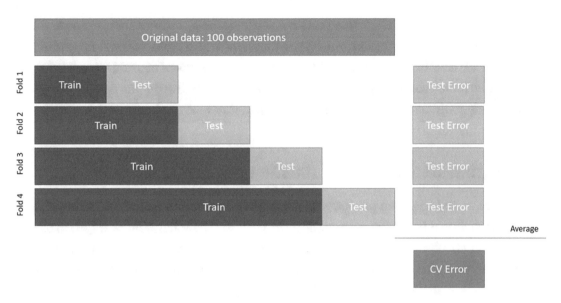

Figure 2-3. *Time series cross-validation in Python*

You can apply time series cross-validation using the code in Listing 2-10.

Listing 2-10. Time series cross-validation in Python

```
from sklearn.model_selection import TimeSeriesSplit
tscv = TimeSeriesSplit()

errors = []
for train_index, test_index in tscv.split(data):
    train = data.iloc[train_index,:]
    test = data.iloc[test_index,:]
    pred = train['actual'].mean()
    test['forecast'] = pred
    error = mean_squared_error(test['actual'], test['forecast'])
    errors.append(error)

print(np.mean(errors))
```

This method of cross-validation estimates the error to be **194.7**. Again, this error should not be used as a stand-alone, but as a way to compare performances of multiple methods on the same data.

This method can at least be used on time series models, as there are no gaps in the data. However, with this method, you have a bias. For the first fold, you will have much less historical data than for the last fold. This makes the errors of the folds to not be comparable: if they cannot use enough historical data for fitting the model, the errors of the first folds could be considered unfair. This is an inherent weakness of this method that you should keep in mind if you use it.

Rolling Time Series Cross-Validation

An alternative that has been proposed is the **rolling time series split**. It uses the same period for each fold in the cross-validation. You can see the schematic overview of this method in Figure 2-4. This will solve the problem of having unequal errors: they are all trained with the same amount of historical data. Yet a problem remains that you cannot base your model on data very far into the past using this method.

But it should only be used in situations where you have a lot of data or where you don't have to look very far into the past for predicting the future.

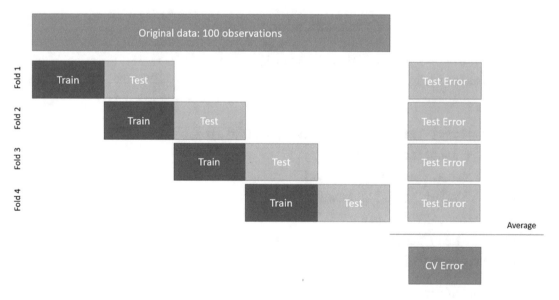

Figure 2-4. *Rolling time series cross-validation in Python*

You can use the code in Listing 2-11 to execute the rolling time series cross-validation on the stock price example.

Listing 2-11. Rolling time series cross-validation in Python

```
from sklearn.model_selection import TimeSeriesSplit
tscv = TimeSeriesSplit(max_train_size = 2)

errors = []
for train_index, test_index in tscv.split(data):
    train = data.iloc[train_index,:]
    test = data.iloc[test_index,:]

    pred = train['actual'].mean()
    test['forecast'] = pred
    error = mean_squared_error(test['actual'], test['forecast'])
    errors.append(error)

print(np.mean(errors))
```

This method of model evaluation estimates the error to be **174.0.** You have now seen three cross-validation errors that are all different. Yet each of them uses the same model. This tells you two things: Firstly, it insists on only comparing error estimates using the same metric, using the same dataset. Secondly, you have observed a significantly lower error in the rolling time series split. This may be a hint that using the model with very little historical data (rolling) could be the more performant model in this case. Yet to be confident of that conclusion, you would need to study the results a bit further. A great starting point would be to add multiple models into the benchmark. This is left for later, as you will discover numerous models throughout this book.

Backtesting

As a last strategy, I want to mention backtesting. Backtesting is a term that is not much practiced in data science and machine learning. Yet it is the go-to model validation technique in stock trading and finance. Backtesting builds on the same principles as model evaluation.

There is a fundamental difference between backtesting and model evaluation using metrics. In backtesting, rather than measuring the accuracy of a forecasting model, you measure the result of a (stock trading) strategy. Rather than trying to forecast the future prices of stocks, you define at which event or trigger you want to buy or sell. For example, you could define a strategy stating that you will buy a given stock at price x and sell at price y.

To evaluate the performances of this strategy, you run the strategy on historical data of stock prices: what would have happened if you had sold and bought stocks at the prices you indicated. Then you evaluate how much profit you would have obtained.

Although backtesting is applied to trading strategies rather than to forecasts, I found it interesting to list it in this chapter, as an alternative to model evaluation in forecasting.

There are several Python libraries for finance, which propose backtesting solutions. A very simple-to-use example is **fastquant**, available on GitHub over here: `https://github.com/enzoampil/fastquant`.

Which Strategy to Use for Safe Forecasts?

After we've seen multiple approaches for forecasting model evaluation, you may wonder which solution you should choose. I'd argue that you do not have to choose. As a forecaster or modeler, your most important task is being confident in the evaluation of your model. A reliable evaluation strategy is the key to success! So rather than choosing one strategy, I'd advise you to use a combination of strategies that makes you feel the most confident about your predictions.

A **combined strategy** that you can use is to train and tune your models using cross-validation on the training data. As a next step, you can measure predictive errors on the validation data: the model that has the best error on the validation data will be your preferred model. As a last verification (you can never be sure enough), you make a prediction on the test data with the selected model, and you make sure that the error is the same as the error observed on the validation data. If this is the case, you can be confident that your model should be delivering the same range of errors on future data.

As you do more predictive forecasts, you will develop a feel for which evaluation metrics are important to consider. Just remember that the goal of model evaluation is twofold: on one hand, you use it to improve model performance, and on the other hand, you use it for objective performance measures.

Be aware that sometimes, predictive modelers are focused so much on the performance improvement part that they start tweaking their evaluation methods to improve their performance scores. This is to be avoided: you want to be as confident as possible about the future performance of your model. Metrics are one important part of this; putting in place an objective model evaluation strategy is even more important.

Final Considerations on Model Evaluation

Besides looking at metrics, you also need to understand what your model is fitting: Which explanatory variables are taken into account by the model? Do those variables make sense intuitively? Do they fit the business knowledge about your data?

This scientific, or explanatory, understanding of your model will give you additional confidence, which is an important complement to the numerical minimization of forecasting errors.

For example, if you predict the stock market, you may obtain a 90% accuracy based on the past. But by reflecting on what you are doing, you may realize that in past data, there has not been a simple market crash present. And you may understand from this that your model will go completely wrong in the case of a market crash, while that is maybe the moment when you are the most at risk ever!

A great book that talks about this is *The Black Swan* in which author Nicholas Taleb insists on the importance of taking into account very rare events when working on stock market trading. A takeaway here is that extreme crises in the history of stock trading have been very influential on trading success, while methods used for day-to-day trading do not (always) take into account such risk management. You may even prefer a model that is worse on a day-to-day basis, but better in predicting huge crashes.

In short, you do not always have to obtain the lowest possible Mean Squared Error! Sometimes, stable performances are the most important. Other times, you may want to avoid any overestimation, while underestimations are not a problem (e.g., in a restaurant, it may be worse to buy too much food due to an overestimated demand forecast, rather than too little). To get to this level of understanding, it is important to understand your models deeply. A model is not a black box that makes predictions for you: it is a mathematical formula that you are building. For good results, you need to evaluate the performances of the formula using evaluation criteria that fit your use case: many strategies and metrics are available, but no one size fits all. Only you can know which evaluation method fits the success criteria of your particular problem statement.

Key Takeaways

- Metrics

    - The most suitable metrics for regression problems are R squared (R2), Root Mean Squared Error, and Mean Squared Error.

    - The R2 gives a percentage-like value. The RMSE gives a value on the scale of the actuals. The MSE gives a value on a scale that is difficult to interpret.

    - Metrics should be used for benchmarking different models on one and the same dataset.

- Model evaluation strategies

    - Cross-validation gives you a very reliable error estimate. Adaptations are necessary to make it work for time series.

    - A train, test, and validation split can be used for benchmarking.

    - A combined strategy will give you the safest estimate: use cross-validation on the training data, validation data for model selection, and test data for a last estimate of the error.

- Overfitting models

    - If your model learns too much from the training data and will not generalize into the future, it is overfitting.

    - Overfitting is identified by a good performance on the train data, but a bad performance on the test data.

- Underfitting models

    - If your model does not learn enough from the training data, it is underfitting.

    - Underfitting is determined by a bad performance on the training data.

PART II

Univariate Time Series Models

CHAPTER 3

The AR Model

In this chapter, you will discover the AR model: the autoregressive model. The AR model is the most basic building block of univariate time series. As you have seen before, univariate time series are a family of models that use only information about the past of the *target variable* to forecast its future, and they do not rely on other *explanatory variables*.

Univariate time series models can be intuitively understood as building blocks: they add up from the simplest model to complex models that combine the different effects described by individual models. The AR model is the simplest model of the univariate time series models.

Throughout the following chapters, you will discover more building blocks to add to this basis. This builds up from the AR model to the SARIMA model. You can see a schematic overview of the building blocks in univariate time series in Table 3-1.

Table 3-1. *The Building Blocks of Univariate Time Series*

Name	Explanation	Chapter
AR	Autoregression	3
MA	Moving Average	4
ARMA	Combination of AR and MA models	5
ARIMA	Adding differencing (I) to the ARMA model	6
SARIMA	Adding seasonality (S) to the ARIMA model	7
SARIMAX	Adding external variables (X) to the SARIMA model *(note that external variables make the model not univariate anymore)*	8

© Joos Korstanje 2021
J. Korstanje, *Advanced Forecasting with Python*, https://doi.org/10.1007/978-1-4842-7150-6_3

I must give you an alert here. As I stated in Chapter 1, the strong point of this book is that it contains real-life examples and honest and objective model evaluations. I cannot insist enough on the importance of model evaluation, as the goal of any forecasting model should be to obtain the best predictive performance.

With the AR model being the simplest building block, it would be unrealistic to expect great predictive performances on most real-life datasets. Although I could have tweaked the example to fit perfectly or work with a simulated dataset, I prefer to show you the weaknesses of certain models as well.

The AR model is great to start learning about univariate time series, but it is unlikely that you will use an AR model without its brothers and sisters in practice. You would generally work with one of the combined models (SARIMA or SARIMAX) and test which are the building blocks that improve predictive performance on your forecast.

Autocorrelation: The Past Influences the Present

The autoregressive model describes a relationship between the present of a variable and its past. It is therefore suitable for variables in which the past and present values correlate.

As an intuitive example, consider the waiting line at the doctor. Imagine that the doctor has a plan in which each patient has 20 minutes with them. If every patient takes exactly 20 minutes, this works fine. But what if a patient takes a bit more time? An **autocorrelation** could be present if the duration of a consultation has an impact on the duration of the next consultation. So, if the doctor needs to speed up a consultation because the previous consultation took too long, you observe a correlation between past and present. Past values influence future values.

Compute Autocorrelation in Earthquake Counts

Throughout this chapter, you will see examples applied to the well-known **Earthquake dataset**. This dataset is collected by the US National Earthquake Information Center. You can find a copy of the dataset on Kaggle (www.kaggle.com/usgs/earthquake-database).

Note   To get a very quick description of a dataset, you can use **df.describe()** to get quick statistics for each column of your dataframe: count, mean, standard deviation, minimum, maximum, and quantiles. For a more thorough description, you can use the **pandas profiling package**. Examples of both are given in the GitHub of this book.

If you import the data using pandas, you can retrieve a description of the dataframe using the **describe** method, as is shown in Listing 3-1. Attention: This code is best executed in a Jupyter notebook or Jupyter lab. These are great tools for exploratory data analysis. You can install them directly, or you can consider installing Anaconda, which comes with many useful Python and data science tools.

Listing 3-1. Describing a dataframe

```
import pandas as pd

# Import the dataframe
eq = pd.read_csv('Ch03_Earthquake_database.csv')

# Describe the dataframe
eq.describe()
```

Feel free to scroll through this descriptive table to get a better understanding of what the data is about. If you want a more detailed description of the data, you can use the **pandas profiling package**, which automatically creates a very detailed description of the variables of a dataframe. You can use Listing 3-2 to get a **profile report**.

Listing 3-2. Profiling a dataframe

```
# Import the pandas profiling package
from pandas_profiling import ProfileReport

# Get the pandas profiling report
eq.profile_report()
```

The task that you'll be performing is forecasting the **number of strong earthquakes per year**. The dataset is currently not in the right format to do this, as it has one line per earthquake and not one line per year. To prepare the data for further analysis, you will need to aggregate the data using Listing 3-3.

Listing 3-3. Convert the earthquake data to the yearly number of earthquakes

```
import matplotlib.pyplot as plt

# Convert years to dates
eq['year'] = pd.to_datetime(eq['Date']).dt.year

# Filter on earthquakes with magnitude of 7 or higher
eq = eq[eq['Magnitude'] >= 7]

# Compute a count of earthquakes per year
earthquakes_per_year = eq.groupby('year').count()

# Remove erroneous values for year
earthquakes_per_year = earthquakes_per_year.iloc[1:-2, 0]

# Make a plot of earthquakes per year
ax = earthquakes_per_year.plot()
ax.set_ylabel("Number of Earthquakes")
plt.show()
```

This code should give you the graph displayed in Figure 3-1. You can see that there might be some relationship between years. One year it's higher and one year it's lower. This could be due to a negative autocorrelation. Remember that a negative correlation implies that a higher value in one variable corresponds to a lower value in the other variable. For autocorrelation, this means that a higher value this year corresponds with a lower value next year.

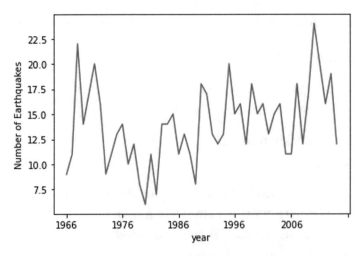

Figure 3-1. *Evaluation of the number of earthquakes per year*

Now that you obtained the number of earthquakes per year, you should **check for autocorrelation numerically**. You need to find out whether there is a correlation between the number of earthquakes in any given year and the year before it.

To do this, you can use the same method for computing correlation as seen in Chapter 2. Correlation is computed pairwise, on two columns. But for now, you just have one column (the data per year). You'll need to add a column in the data containing the data for a previous year. This can be obtained by applying a **shift** to the original data and concatenating the shifted data with the original data. Listing 3-4 shows you how to do this. Figure 3-2 shows you how the data has shifted 1 year back.

Listing 3-4. Plotting the shifted data

```python
shifts = pd.DataFrame(
    {
        'this year': earthquakes_per_year,
        'past year': earthquakes_per_year.shift(1)
    }
)

ax = shifts.plot()
ax.set_ylabel('Number of Earthquakes')
plt.show()
```

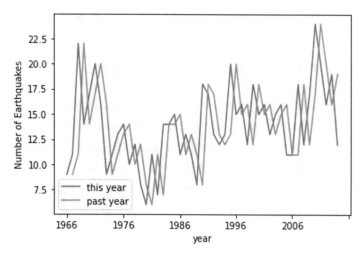

Figure 3-2. *Shifted data*

This works well, except for one thing. As you can see in the data, there are observations from 1966 to 2014. This means that you don't have a value for the past year of 1966: this introduces a **nan value** (a **missing value**). This type of border effect is very common in time series analysis, and the best solution, in this case, is to leave the year 1966 out of consideration. You can delete all rows with missing data using Listing 3-5.

Listing 3-5. Drop missing data

```
shifts = shifts.dropna()
```

The next step is to compute the Pearson correlation coefficient using the code that you've seen in Chapter 2. In case you forgot, you can find it in Listing 3-6. And you can see the result in Figure 3-3.

Listing 3-6. Compute a correlation matrix for the shifts dataframe

```
shifts.corr()
```

	this year	past year
this year	1.000000	0.313667
past year	0.313667	1.000000

Figure 3-3. *Correlation matrix*

In the following paragraph, you will discover how to interpret this autocorrelation.

Positive and Negative Autocorrelation

Just like "regular" correlation, autocorrelation can be positive or negative. Positive autocorrelation means that a high value now will likely give a high value in the next period. This can, for example, be observed in stock trading: as soon as many people want to buy a stock, its price goes up. This positive trend makes people want to buy this stock even more as it has positive results. The more people buy the stock, the more it goes up and the more people may want to buy it.

A positive correlation also works on downtrends. If today's stock value is low, then it is likely that tomorrow's value would be even lower, as people start selling. When a lot of people sell, the value drops, and even more people will want to sell. This is also a case of positive autocorrelation as the past and the present go in the same direction. If the past is low, the present is low; and if the past is high, the present is high.

Negative autocorrelation exists if two trends are oppositive. This is the case in the doctor's consultation durations example. If one consultation takes longer, the next one will take shorter. If one consultation takes less time, the doctor may take a bit more time on the next one.

If you have applied the code in Listing 3-6, you have observed a correlation of around **0.31** between current year and past year. This correlation is positive rather than negative as was expected based on the graph. This correlation is of medium strength. It seems that the correlation has captured the trend in the data. If you look at the graph, you can see that there is a trend in the number of earthquakes: a bit higher in the earliest years, then a bit lower for a certain period, and in the last years a bit higher again. This is a problem, as the trend is not the relationship that you want to capture here.

Stationarity and the ADF Test

The problem of having a trend in your data is general in univariate time series modeling. The **stationarity** of a time series means that a time series does not have a (long-term) trend: it is stable around the same average. If not, you say that a time series is **non-stationary**.

AR models can theoretically have a trend coefficient in the model, yet since stationarity is an important concept in the general theory of time series, it is better to learn how to deal with it right away. A lot of models can only work on stationary time series.

A time series that is strongly growing or diminishing over time is obvious to spot. But sometimes it is difficult to tell whether a time series is stationary. This is where the **Augmented Dickey Fuller (ADF) test** comes in useful. The Augmented Dickey Fuller test is a hypothesis test that allows you to test whether a time series is stationary. It is applied as shown in Listing 3-7.

Listing 3-7. Augmented Dickey Fuller test

```
from statsmodels.tsa.stattools import adfuller
result = adfuller(earthquakes_per_year.dropna())
print(result)

pvalue = result[1]
if pvalue < 0.05:
    print('stationary')
else:
    print('not stationary')
```

In the earthquake data case, you will see a **p-value** that is smaller than **0.05** (the reference value), and this means that the series is theoretically stationary: you don't have to change anything to apply the AR model.

Note   If you're not familiar with hypothesis testing, you should know that the p-value is the conclusive value for a hypothesis test. In a hypothesis test, you try to prove an alternative hypothesis against a null hypothesis. The p-value indicates the probability that you would observe the data that you have observed if the null hypothesis were true. If this probability is low, you conclude that the null hypothesis must be wrong and therefore the alternative must be true. The reference value for the p-value is generally 0.05 but may differ for certain applications.

Differencing a Time Series

Even though the ADF test tells you that the data is stationary, you have seen that the autocorrelation is positive where you would expect a negative one. The hypothesis was that this could be caused by a trend.

You now have two opposing indicators. One tells you that the data is stationary, and the other one tells you that there is a trend. To be safe, it is better to remove the trend anyway. This can be done by **differencing**.

Differencing means that rather than modeling the original values of the time series, you model the differences between each value and the next. Even though there is a trend in the actual data, there will probably not be a trend in the differenced data. To confirm, you can do an ADF test again. If it is still not good, you can difference the differenced time series again, until you obtain a correct result.

Listing 3-8 shows you how you can easily difference your data in pandas.

Listing 3-8. Differencing in pandas

```
# Difference the data
differenced_data = earthquakes_per_year.diff().dropna()

# Plot the differenced data
ax = differenced_data.plot()
ax.set_ylabel('Differenced number of Earthquakes')
plt.show()
```

You can see what the differenced data looks like in Figure 3-4.

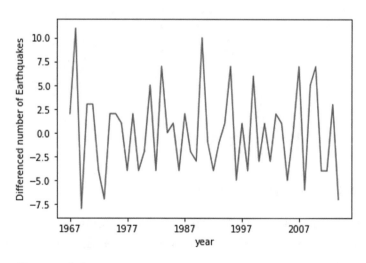

Figure 3-4. *Differenced data*

Now to see whether this has switched your autocorrelation to the right direction, you can compute the correlation coefficient as you did before. The code for this can be seen in Listing 3-9. You can see what this shifted and differenced data looks like in Figure 3-5. The correlation matrix (Figure 3-6) shows the correlation coefficient between the shifted differenced data and the nonshifted differenced data.

Listing 3-9. Autocorrelation of the differenced data

```
shifts_diff = pd.DataFrame(
    {
        'this year': differenced_data,
        'past year': differenced_data.shift(1)
    }
)

ax = shifts_diff.plot()
ax.set_ylabel('Differenced number of Earthquakes')
plt.show()

shifts_diff.corr()
```

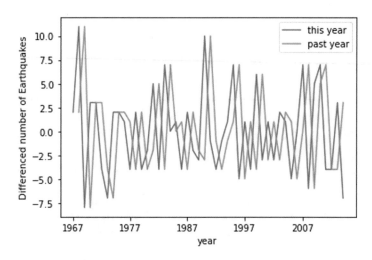

Figure 3-5. *Shifted and differenced data*

You now obtain a correlation coefficient of **-0.37**, as can be seen in Figure 3-6: you have successfully removed the trend from the data. Your correlation is now the negative correlation that you expected. If there are more earthquakes one year, then you generally have fewer earthquakes the next year.

	this year	past year
this year	1.000000	-0.376859
past year	-0.376859	1.000000

Figure 3-6. *Autocorrelation on differenced data*

Lags in Autocorrelation

Besides the direction of the autocorrelation, the notion of **lag** is important. This notion is applicable only in autocorrelation and not in "regular" correlation.

The number of lags of an autocorrelation means the number of steps back in time that have an impact on the present value. In many cases, if autocorrelation is present, it is not just the previous time step that has an impact, but it can be many time steps.

Applied to the doctor's consultation durations example, you could imagine that a doctor has a very long delay in one consultation and that they need to speed up for the next three consultations. This would mean that the third "speedy" consultation is still impacted by the delayed one. So the lag of autocorrelation would be at least three: three steps back in time.

The doctor's consultation durations example is very intuitive, as it is very logical that a doctor has a limited amount of time. In the example of earthquakes per year, there is not such a clear logic that explains relationships between past values and current values. This makes it difficult to intuitively identify the number of lags that should be included.

A great tool to investigate autocorrelation on multiple lags at the same time is to use the autocorrelation plot from the **statsmodels package**. This code is shown in Listing 3-10. **ACF** is short for **autocorrelation function**: the autocorrelation as a function of lag.

Note that the choice for 20 lags is arbitrary: it is for plotting purposes only. Twenty lags mean 20 years back in time, and this should be enough to show relevant autocorrelations, but 40 or even more would be an acceptable option too.

Listing 3-10. Autocorrelation of the differenced data

```
from statsmodels.graphics.tsaplots import plot_acf
import matplotlib.pyplot as plt

plot_acf(differenced_data, lags=20)
plt.show()
```

You will obtain the plot in Figure 3-7. This plot shows which autocorrelations are important to retain. On the y-axis, you see the **correlation coefficient** between the non-lagged data (original data) and the lagged data. The correlation coefficient is between -1 and 1. On the x-axis, you see the **number of lags** applied, in this case from 0 to 20. The first correlation is 1, as this is the data with 0 lag: the original data.

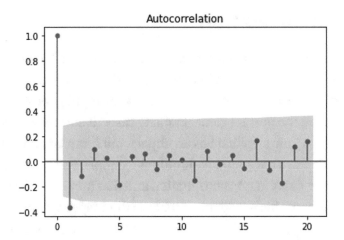

Figure 3-7. *Autocorrelation plot*

In this plot, you should look for **autocorrelation** values that are at least higher than 0.2 or lower than -0.2. Anything below that can be considered noise. Using 0.3 or 0.4 as a minimum value is also possible: there is no strict guideline in interpreting the correlation coefficient, as in some domains data are noisier by nature. For example, in social studies, when working with measurements from questionnaire studies, you'll often see much more noise than when working with precise physics measurements.

The blue area in the graph allows you to detect visually which autocorrelations are just noise: if the spike goes outside of the blue area, there is a significant autocorrelation, but if the spike stays within the blue area, the observed autocorrelation is insignificant. You can generally expect the autocorrelation to be higher in lags that are closer to the present and diminish toward further-away moments.

In the graph in Figure 3-7, you observe a first spike (at lag 0), which is very strong. This is normal, as it is the original data. The correlation between a variable and itself is always 1 (100%). The second spike is the correlation with lag 1. This is the autocorrelation that you have computed before using the correlation matrix. The correlation coefficient was -0.37, and you see this again in this graph.

The remaining autocorrelations are less strong. This means that to know the number of earthquakes today, we should mainly look at the values of one lag back; in the current example, this means the number of earthquakes last year. If there were many earthquakes last year, we can expect fewer earthquakes this year: this is indicated by the negative autocorrelation for lag 1.

For the sake of understanding how to interpret further lags, imagine a different (hypothetical) ACF plot with only one positive spike at lag 5 and the other lags being all 0. This would mean that to understand this year's number of earthquakes, you need to look what happened 5 years ago: if there were many earthquakes five years ago, there will be many earthquakes this year. What happened in between is not relevant: since there is no autocorrelation with the previous four years, the number of earthquakes in the previous four years will not help you determine the number of earthquakes this year. This could, for example, be a case of a seasonality over 5-year periods.

Partial Autocorrelation

The **partial autocorrelation plot** is another plot that you should also look at. The difference between autocorrelation and partial autocorrelation is that partial autocorrelation makes sure that any correlation is not counted multiple times for multiple lags. Therefore, the partial autocorrelation for each lag is **additional autocorrelation** to each inferior lag. More mathematically correct is to say that partial autocorrelation is **autocorrelation conditional on earlier lags.**

$$PAC(y_i, y_{i-h}) = \frac{cov(y_i, y_{i-h}|y_{i-1}, ..., y_{i-h})}{\sqrt{var(y_i|y_{i-1}, ..., y_{i-h+1}) * var(y_{i-h}|y_{i-1}, ..., y_{i-h+1})}}$$

This has an added value, as the correlations in the autocorrelation plot may be redundant between one another.

The idea is easier to understand on a hypothetical example. Imagine a hypothetical case in which the values for the present, lag 1 and lag 5, are exactly the same. An autocorrelation would tell you that there is perfect autocorrelation with both lag 1 and lag 5. Partial autocorrelation would tell you that there is perfect autocorrelation with lag 1, but it would not show autocorrelation for lag 5. As you should always try to make models with the lowest number of variables necessary (often referred to as **Occam's razor** or the **parsimony principle**), it would be better to include only lag 1 and not lag 5 in this case.

You can compute the partial autocorrelation plot using statsmodels, as is shown in Listing 3-11. **PACF** is short for **partial autocorrelation function**: the autocorrelation as a function of lag.

Listing 3-11. Partial autocorrelation of the differenced data

```
from statsmodels.graphics.tsaplots import plot_pacf
plot_pacf(differenced_data, lags = 20)
plt.show()
```

This code will give you the plot shown in Figure 3-8. As you can see based on the blue background shading in the graph, the PACF shows the first and the second lag outside of the shaded area. This means that it would be interesting to also include the second lag in the AR model.

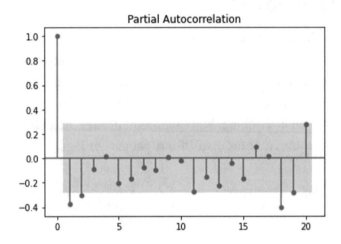

Figure 3-8. *Partial autocorrelation*

How Many Lags to Include?

Now the big question in time series analysis is always how many lags to include. This is called the **order of the time series**. The notation is **AR(1)** for **order 1** and **AR(p)** for an **order p**.

The order is up to you to decide. Theoretically speaking, you can base your order on the PACF graph. The theory tells you to take the number of lags before obtaining an autocorrelation of 0. All the other lags should be 0.

In the theory, you often see great graphs where the first spike is very high and the rest is equal to zero. In those cases, the choice is easy: you are working with a very "pure" example of AR(1). Another common case is when your autocorrelation starts high and slowly diminishes to zero. In this case, you should use all of the lags where the PACF is not yet zero.

Yet, in practice, it is not always this simple. Remember the famous saying "all models are wrong, but some are useful." It is very rare to find cases that perfectly fit an AR model. In general, the autoregression process can help to explain part of the variation in a variable, but not all of it.

In the earthquake example, there is still a bit of partial autocorrelation on further lags, and this goes on until very far lags. There is even a spike of PACF on lag 18. This could mean many things: maybe there really is autocorrelation with lags far away. Or maybe a hidden process is underlying that the AR model is not capturing well.

In practice, you will try to select the number of lags that gives your model the **best predictive performance**. Best predictive performance is often not defined by looking at autocorrelation graphs: those graphs give you a theoretical estimate. Yet predictive performance is best defined by doing **model evaluation and benchmarking**, using the techniques that you have seen in Chapter 2. Later in this chapter, you will see how to use model evaluation to choose a performant order for the AR model. But before getting into that, it is time to go deeper into the exact model definition of the AR model.

AR Model Definition

Until now, you have seen many aspects of the autoregressive model intuitively. As seen before, autocorrelations show for each lag whether there is a correlation with the present. The AR model uses those correlations to make a predictive model.

But the most important part is still missing: the mathematical definition of the AR model. To make a prediction based on autocorrelation, you need to express the future values of the target variable as a function of the lagged variables. For the AR model, this gives the following model definition:

$$X_t = \sum_{i=1}^{p} \varphi_i X_{t-i} + \varepsilon_t$$

In this formula, X_t is the current value, and it is computed as the sum of each lagged value X_{t-i} multiplied by a coefficient φ_i for this specific lag. The error ε_t at the end is random noise that you cannot predict, but you can estimate.

Estimating the AR Using Yule-Walker Equations

You have now seen **the definition of the AR model**. That is, you have defined its form. However, you can't use it as long as you don't have the values to plug into the formula. The next step is to **fit the model**: you need to find the optimal values for each coefficient to be able to use this. This is also called **estimating the coefficients**.

Once you have values for the parameters, it is relatively easy to use the model: just plug in the lagged values. But as in any machine learning theory, the difficult part is always in estimating the model.

In the formula, you see that the past values (X_{t-i}) are multiplied by a coefficient (φ_i). The sum of this will give you the current value X_t. When starting to fit the model, you already know the past values and the current value. The only thing you don't know is the phis (φ_i). The phis are what needs to be found mathematically to define the model. Once you estimate those phis, you will be able to compute tomorrow's value of X_t using the estimated phis and the known values of today and before.

For the AR model, different methods have been found to estimate the coefficients (the phis), but the Yule-Walker method is generally accepted as the most appropriate method.

The Yule-Walker Method

The **Yule-Walker method** consists of a set of equations:

$$\gamma_m = \sum_{k=1}^{p} \varphi_k \gamma_{m-k} + \sigma_\varepsilon^2 \delta_{m,0}$$

In this formula, γ_m is the **autocovariance function** of X_t. m is the number of lags from 0 to p (p is the maximum lag). There are therefore p + 1 equations. σ_ε^2 is the standard deviation of the errors.

$\delta_{m,0}$ is the **Kronecker delta function**: it returns 1 if both values are equal (so if m is equal to 0) and 0 if m is not 0. You can see that this means that everything after the + sign is equal to σ_ε^2 if m is 0 and the whole is equal to 0 otherwise. Therefore, there is only one equation that has the part after the +, while the others (from m = 1 to m = p) do not have this part.

The clue to solving this is to start with only the equations from m = 1 to m = p (p equations) written in **matrix format**:

$$
\begin{pmatrix} \gamma_1 \\ \gamma_2 \\ \gamma_3 \\ \cdots \\ \gamma_p \end{pmatrix} = \begin{pmatrix} \gamma_0 & \gamma_{-1} & \gamma_{-2} & \cdots \\ \gamma_1 & \gamma_0 & \gamma_{-1} & \cdots \\ \gamma_2 & \gamma_1 & \gamma_0 & \cdots \\ \cdots & \cdots & \cdots & \cdots \\ \gamma_{p-1} & \gamma_{p-2} & \gamma_{p-3} & \cdots \end{pmatrix} \begin{pmatrix} \varphi_1 \\ \varphi_2 \\ \varphi_3 \\ \cdots \\ \varphi_p \end{pmatrix}
$$

This can be written in **matrix notation** as

$$
\gamma = \Gamma^{-1}\widehat{\Phi}
$$

To obtain the values for phis, you can use the **Ordinary Least Squares** (OLS) method. The Ordinary Least Squares method is a matrix computation that is applied widely throughout statistics to solve problems like this one. It allows estimating coefficients in cases where you have a matrix with historical data and a vector with historical outcomes. The solution to finding the best estimates for the phis is the following:

$$
\widehat{\Phi} = \left(\Gamma^{\mathrm{T}}\Gamma\right)^{-1}\Gamma^{\mathrm{T}}\gamma
$$

You will see the OLS method also in Chapter 11 and other chapters throughout this book.

Once you've applied OLS, you will have obtained the estimates for phis, and you can then compute the sigma by using the estimated phis for solving:

$$
\gamma_0 = \sum_{k=1}^{p} \varphi_k \gamma_{m-k} + \sigma_\varepsilon^2
$$

Now you'll start fitting the AR model in Python, using Listing 3-12. Fitting the model in this case means to identify the best possible values for the phis. As you're working with real data and not with a simulated sample, note that you'll be working toward the best fit possible of the AR model throughout this example.

Listing 3-12. Estimate Yule-Walker AR coefficients with order 3

```
from statsmodels.regression.linear_model import yule_walker
coefficients, sigma = yule_walker(differenced_data, order = 3)
```

```
print('coefficients: ', -coefficients)
print('sigma: ', sigma)
```

The **coefficients** that you obtain with this are the coefficients for each of the lagged variables. In this case, the order is 3, so that will need to be applied for the three last values to compute the new value. You can make a forecast by computing the next **steps**, based on the current coefficients. The code for this is shown in Listing 3-13. Attention: The more steps forward you want to forecast, the more difficult it becomes – your error will likely increase with the number of steps.

Listing 3-13. Make a forecast with the AR coefficients

```
coefficients, sigma = yule_walker(differenced_data, order = 3)

# Make a list of differenced values
val_list = list(differenced_data)
# Reverse the list so that the order corresponds with the order of the
  coefficients
val_list.reverse()
# Define the number of years to predict
n_steps = 10

# For each year to predict
for i in range(n_steps):

    # Compute the new value as the sum of lagged values multiplied by their
      corresponding coefficient
    new_val = 0
    for j in range(len(coefficients)):

        new_val += coefficients[j] * val_list[j]

    # Insert the new value at the beginning of the list
    val_list.insert(0, new_val)

# Redo the reverse to have the order of time
val_list.reverse()

# Add the original first value back into the list and do a cumulative sum
  to undo the differencing
```

```
val_list = [earthquakes_per_year.values[0]] + val_list
new_val_list = pd.Series(val_list).cumsum()

# Plot the newly obtained list
plt.plot(range(1966, 2025), new_val_list)
plt.ylabel('Number of earthquakes')
plt.show()
```

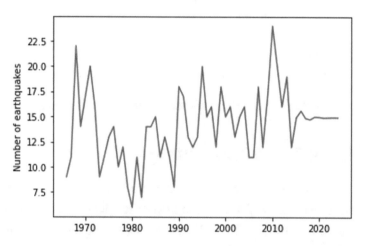

Figure 3-9. *The AR(3) model has not captured anything: flat line forecast*

You will obtain the plot in Figure 3-9. The first values are actual data, and the last ten steps are forecasted data. Unfortunately, you can see that the model has not captured anything interesting: it is just a flat line! But you haven't finished yet. Let's try with an order of 20 this time and see whether anything improves.

Just change the order in the first line, and you will obtain the plot in Figure 3-10. It obtains a very different forecast than the one in Figure 3-9. Your model seems to be learning something much more interesting when you add an order 20: it has captured some variation.

Although this seems to fit well with the data visually, you still need to verify whether the model was correct: did it forecast something close to reality? To do this, you will use some strategies and metrics from Chapter 2 and apply them to this forecast.

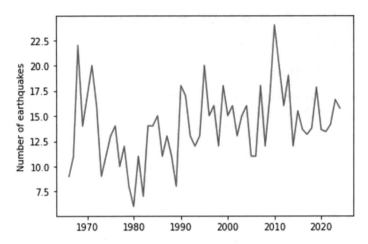

Figure 3-10. *The forecast with order 20 has captured something, but is it correct?*

Train-Test Evaluation and Tuning

In this case, let's apply a train-test split. You will cut off the last 10 years of the dataset and use them as a test set. This allows you to fit your model using the rest of the data (the train set). Once you have fitted on the train set, you can compute the error between the test set and your prediction. The code in Listing 3-14 does just that.

Listing 3-14. Fit the model on a train set and evaluate it on a test set

```
from sklearn.metrics import r2_score

train = list(differenced_data)[:-10]
test = list(earthquakes_per_year)[-10:]

coefficients, sigma = yule_walker(train, order = 3)

# Make a list of differenced values
val_list = list(train)
# Reverse the list so that the order corresponds with the order of the
  coefficients
val_list.reverse()
# Define the number of years to predict
n_steps = 10
```

```python
# For each year to predict
for i in range(n_steps):

    # Compute the new value as the sum of lagged values multiplied by their
      corresponding coefficient
    new_val = 0
    for j in range(len(coefficients)):

        new_val += coefficients[j] * val_list[j]

    # Insert the new value at the beginning of the list
    val_list.insert(0, new_val)

# Redo the reverso to have the order of time
val_list.reverse()

# Add the original first value back into the list and do a cumulative sum
  to undo the differencing
val_list = [earthquakes_per_year[0]] + val_list
new_val_list = pd.Series(val_list).cumsum()

# Plot the newly obtained list
validation = pd.DataFrame({
    'original': earthquakes_per_year.reset_index(drop=True),
    'pred': new_val_list })

print('Test R2:', r2_score(validation.iloc[-10:, 0], validation.iloc[-10:, 1]))

# Plot the newly obtained list
plt.plot(range(1966, 2015), validation)
plt.legend(validation.columns)
plt.ylabel('Number of earthquakes')
plt.show()
```

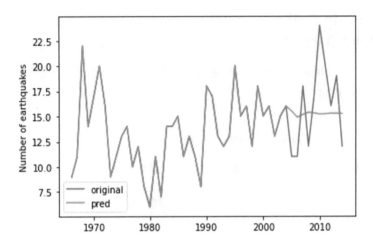

Figure 3-11. *Evaluation plot of the AR model with order 3*

Unfortunately, the model with order 3 is quite bad. So bad that the R2 is negative (**-0.04**). You can see from the graph (Figure 3-11) that the model suffers from **underfitting**: it did not capture anything and predicts only an average value into the future. This confirms what you've seen before with order 3.

In the previous example, you have applied the order 20 to see whether that would give a better result. Yet the choice for order 20 is quite random. To find an order that works well, you can apply a **grid search**.

A grid search consists of doing a model evaluation for each value of a **hyperparameter**. Hyperparameters are parameters that are not estimated by the model but chosen by the modeler. The order of the model is an example of this. Other models often have more hyperparameters, which makes choosing hyperparameters a nontrivial decision.

A grid search for one parameter is very simple to code: you simply fit the model for each possible value for the hyperparameter, and you select the best-performing one. You can use the example in Listing 3-15.

Listing 3-15. Apply a grid search to find the order that gives the best R2 score on the test data

```
def evaluate(order):
    train = list(differenced_data)[:-10]
    test = list(earthquakes_per_year)[-10:]

    coefficients, sigma = yule_walker(train, order = order)
```

```python
    # Make a list of differenced values
    val_list = list(train)
    # Reverse the list to corresponds with the order of coefs
    val_list.reverse()
    # Define the number of years to predict
    n_steps = 10

    # For each year to predict
    for i in range(n_steps):

        # Compute the new value
        new_val = 0
        for j in range(len(coefficients)):
            new_val += coefficients[j] * val_list[j]

        # Insert the new value at the beginning of the list
        val_list.insert(0, new_val)

    # Redo the reverse to have the order of time
    val_list.reverse()

    # Undo the differencing with a cumsum
    val_list = [earthquakes_per_year[0]] + val_list
    new_val_list = pd.Series(val_list).cumsum()

    # Plot the newly obtained list
    validation = pd.DataFrame({
        'original': earthquakes_per_year.reset_index(drop=True),
        'pred': new_val_list })

    return r2_score(validation.iloc[-10:, 0], validation.iloc[-10:, 1])

# For each order between 1 and 30, fit and evaluate the model
orders = []
r2scores = []
for order in range(1, 31):
    orders.append(order)
    r2scores.append(evaluate(order))
```

```
# Create a results data frame
results =pd.DataFrame({'orders': orders,
                       'scores': r2scores})
```

```
# Show the order with best R2 score
results[results['scores'] == results.max()['scores']]
```

This gives the highest R2 score for order 19, with an R2 score of **0.13**. This means that the AR(19) model explains about 13% of the variation on the test data.

You can reuse the code in Listing 3-13 to recreate the graph with the forecast using order 19. Just change the order in the line that fits the Yule-Walker coefficients. The graph is shown in Figure 3-12.

This code will also make a plot for the model with order 19 and observe that, although not perfect, the model indeed fits a bit better than the model with order 3.

The AR model is one of the basic building blocks of univariate time series. It can be used as a **stand-alone** model only in very rare and specific cases. In this real-life example of predicting earthquakes, the R2 score of 0.13 is not very good: you can surely do much better by including other building blocks of univariate time series. You will discover these building blocks throughout the next chapters.

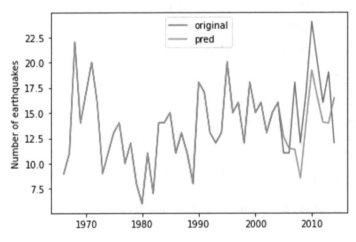

Figure 3-12. *Forecast on the test set for the AR model with order 19*

Key Takeaways

- The AR model predicts the future of a variable by leveraging correlations between a variable's past and present values.

- Autocorrelation is correlation between a time series and its previous values.

- Partial autocorrelation is autocorrelation conditional on earlier lags: it prevents double counting correlations.

- The number of lags to include in the AR model can be based on theory (ACF and PACF plots) or can be determined by a grid search.

- A grid search consists of doing a model evaluation for each value of the hyperparameters of a model. This is an optimization method for the choice of hyperparameters.

- Hyperparameters are different from model coefficients. Model coefficients are optimized by a model, while hyperparameters are chosen by the modeler. Yet the model can use optimization techniques like a grid search to find the best hyperparameters.

- Yule-Walker equations are used to fit the AR model. Fitting the model means finding the coefficients of the model.

CHAPTER 4

The MA Model

The MA model, short for the **Moving Average model**, is the second important building block in univariate time series (see Table 4-1). Like the AR model, it is a building block that is more often used as a part of more complex time series, but it can also be used as a stand-alone.

Table 4-1. *The Building Blocks of Univariate Time Series*

Name	Explanation	Chapter
AR	Autoregression	3
MA	Moving Average	4
ARMA	Combination of AR and MA models	5
ARIMA	Adding differencing (I) to the ARMA model	6
SARIMA	Adding seasonality (S) to the ARIMA model	7
SARIMAX	Adding external variables (X) to the SARIMA model *(note that external variables make the model not univariate anymore)*	8

Throughout this chapter, you will see the MA model applied to a forecasting example. Please note that, as in the previous chapter, the model is optimized as **a stand-alone model.** When applying univariate time series in practice, you will generally use the SARIMA or SARIMAX model directly: after all, the MA model is a component of this model. Yet there is added value in seeing the MA model applied separately, as it is important to understand each building block of the SARIMAX model separately.

© Joos Korstanje 2021
J. Korstanje, *Advanced Forecasting with Python*, https://doi.org/10.1007/978-1-4842-7150-6_4

The Model Definition

The mathematical model underlying the MA model is quite like that of the AR model in form but very different in intuitive understanding. Where the AR model looks at lagged values of the target variable to predict the future, the MA model looks at the **model error of past predictions**. The model definition is as follows:

$$X_t = \mu + \varepsilon_t + \theta_1 \varepsilon_{t-1} + \ldots + \theta_q \varepsilon_{t-q}$$

In this equation, **Xt** are the current values (unlagged), μ (mu) is the average of the series, ε (epsilon) is the error of the model, and θ (thetas) are the coefficients to multiply with each of the past values.

As you can see, the model has only past errors in it and no past values. This means that to predict the future, you don't look at what happened in the past, but only at how wrong you were in the past!

This may seem strange, but it has some logic in it. Imagine a case in which you try to make a prediction based on the past. At some points, the process differs strongly from what you expected, and you do not understand why. Let's say you had a huge unexpected and temporary drop in stock prices at some point.

Both an AR model and an MA model would be able to take this drop into account over time, but in a different way. The AR model would model it as a past low value. The MA model would model it as a large error, which can be interpreted as a large impulse. In this case, it makes sense to see the error as the impulse rather than the value itself, because the fact that the impulse was unexpected is what is going to be a driver for the stock prices the next days as stock traders will be reacting to this unexpected impulse.

In this way, *the model error in the past influences the future, and past errors can help you to make a forecast*. This is the idea behind the MA model.

Note   The term moving average is also used for taking the average of the most recent values. In time series, this is often used for smoothing a time series. This has nothing to do with the MA model.

Fitting the MA Model

Fitting the MA model is more complicated than the AR model. From a high level, the MA model seems relatively comparable to an AR model: MA depends on past errors, whereas the AR model depends on past values.

Yet there is a big difference here. When fitting the MA model, you do not yet have the past errors of the model. Since you have not yet defined the model, you cannot estimate the error of the model. And you need the error of the model to estimate the coefficients.

There is some sort of a circular dependency here, which makes it impossible to find a straightforward method to fit the model at once. Yet different techniques can be used for fitting an MA model, of which the most common is the Nonlinear Least Squares. It assumes that the error at time 0 is negligible. The thetas can then be estimated using the following formula:

$$\hat{\theta}_{nls} = \operatorname*{argmin}_{\theta} \sum_{t=2}^{T} \left(x_t - \theta \sum_{k=0}^{t-2} (-\theta)^k \varphi_{t-1-k} \right)^2$$

Since this formula uses an **argmin**, it is a numerical search for the optimal values for thetas. This means that fitting the model is simply done by iteratively searching through different values for thetas and identifying which values for thetas lead to the lowest value for the rest of the formula, that is, the lowest of the following:

$$\sum_{t=2}^{T} \left(x_t - \theta \sum_{k=0}^{t-2} (-\theta)^k \varphi_{t-1-k} \right)^2.$$

An example of such a minimization can be found in Figure 4-1. The figure shows a curve of error, which depends on the parameters. The goal is to find the parameters that give you the lowest model error. This is not an easy task, as you generally have no clue what those parameters are.

I won't go into more depth in the optimization methods that can be used for this minimization task, as that would be out of scope for this book. But you should know that there are many algorithms that do such optimization tasks.

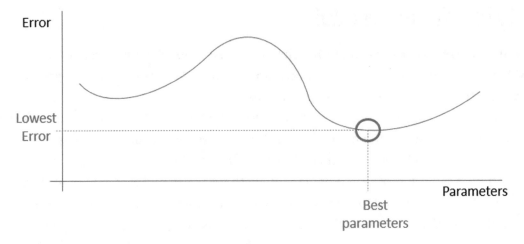

Figure 4-1. *Parameters that minimize an error*

Stationarity

Important to note is that the MA model has an absolute need for stationarity. Where the AR model can adapt to non-stationary situations, the MA model cannot. The Augmented Dickey Fuller test and differencing are notions that have been covered in Chapter 3 and can be used here as well.

Choosing Between an AR and an MA Model

Just like the AR model, the lag of the MA model can be decided by looking at the PACF plot (the partial autocorrelation function). But there is more use to the PACF curve than just selecting the best theoretical lag.

The PACF plot can also help you to choose the correct type of model. For instance, you have now seen the AR and the MA model, but you have no idea which one to use. The type of autocorrelation can be a factor in this decision.

The following are indicators for the type of model to use:

- Autoregression (typical PACF plot in Figure 4-2):

  - Autocorrelation decays to 0.

  - Autocorrelation decays exponentially.

  - Autocorrelation alternates between positive and negative and decays to 0.

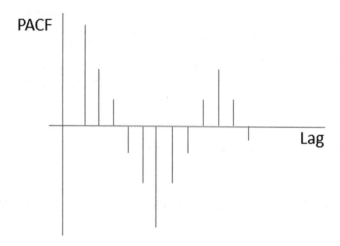

Figure 4-2. *Typical AR model PACF*

- Moving average (typical ACF plot in Figure 4-3):

  - Many values of (almost) zero and a few spikes

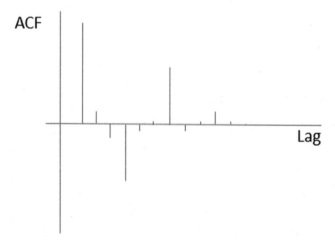

Figure 4-3. *Typical MA model ACF*

Application of the MA Model

Let's now apply this to an example. The data that you will use for this example are Microsoft stock closing prices. You can get those data through the Yahoo Finance Python package as you can see in Listing 4-1.

Note   The Yahoo Finance package is a Python package that allows downloading stock data from the Yahoo Finance website.

Listing 4-1. Importing stock price data using the Yahoo Finance package

```
from pandas_datareader import data as pdr
import yfinance

data = pdr.get_data_yahoo('MSFT', start='2019-01-01', end='2019-12-31')
data = data['Close']
```

As you know, it is important to get a first idea of the data, so let's make a plot of the closing prices. This is done in Listing 4-2.

Listing 4-2. Plotting the stock price data

```
import matplotlib.pyplot as plt
ax = data.plot()
ax.set_ylabel("Stock Price")
plt.show()
```

This will output the graph in Figure 4-4.

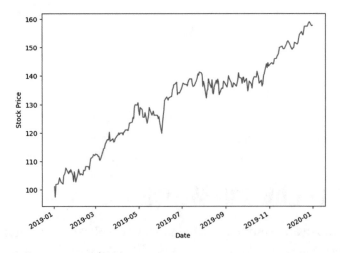

Figure 4-4. *Plot of the original data*

In this graph, you can clearly see an upward trend. It is not even necessary to apply an Augmented Dickey Fuller (ADF) test here to see that this data is **not stationary**. Since the MA model cannot function without stationarity, let's apply the go-to solution: differencing. This is done in Listing 4-3 and will show the plot in Figure 4-5.

Listing 4-3. Computing the differenced data and plotting it

```
# Need to difference
data = data.diff().dropna()
ax = data.plot()
ax.set_ylabel("Daily Difference in Stock Price")
plt.show()
```

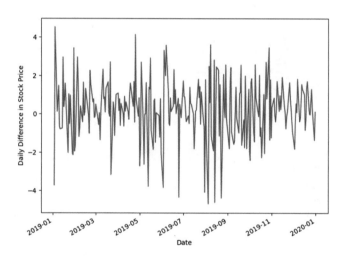

Figure 4-5. *Plot of the differenced data*

This differenced data seems stationary. For consistency's sake, Listing 4-4 shows how to use an Augmented Dickey Fuller test on these differences, although by visual inspection, there is really no doubt about it: the data is clearly varying around a fixed mean of 0.

Listing 4-4. Applying an ADF test to the differenced data

```
from statsmodels.tsa.stattools import adfuller
result = adfuller(data)
```

```
pvalue = result[1]
if pvalue < 0.05:
    print('stationary')
else:
    print('not stationary')
```

This confirms stationarity of the differenced series. Let's have a look at the autocorrelation and partial autocorrelation functions to see whether there is an obvious choice for the lag using Listing 4-5 to obtain the ACF in Figure 4-6 and the PACF in Figure 4-7.

Listing 4-5. Plotting the autocorrelation function and the partial autocorrelation function

```
from statsmodels.graphics.tsaplots import plot_acf, plot_pacf
plot_acf(data, lags=20)
plot_pacf(data, lags=20)
plt.show()
```

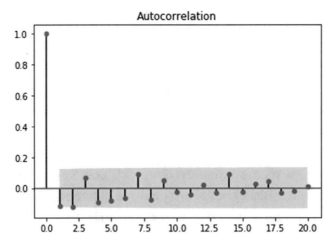

Figure 4-6. *Autocorrelation function*

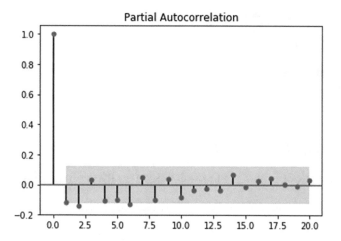

Figure 4-7. *Partial autocorrelation function*

Unfortunately, there is not a very clear pattern in the autocorrelation nor partial autocorrelation that would confirm a choice for the order. You can observe some decay in the partial autocorrelation function, which is a positive sign for using a time series approach. But nothing too obvious on the choice of lag. Let's fit the MA model with order 1 and see what it does.

To fit the MA model in Python, you need to use the ARIMA function from statsmodels. ARIMA is a model that you will see in a next chapter. It is a model that contains multiple building blocks of univariate time series, including AR for the AR model and MA for the MA model. You can specify the order for each of the "smaller" models separately, so by setting everything except MA to 0, you obtain the MA model. This can be done using Listing 4-6.

Listing 4-6. Fitting the MA model and plotting the forecast

```
from sklearn.metrics import r2_score
from statsmodels.tsa.arima.model import ARIMA

# Forecast the first MA(1) model
mod = ARIMA(data.diff().dropna(), order=(0,0,1))
res = mod.fit()
```

```
orig_data = data.diff().dropna()
pred = res.predict()

plt.plot(orig_data)
plt.plot(pred)
plt.show()

print(r2_score(orig_data, pred))
```

This gives you a plot of the fit (Figure 4-8) and an R2 score on the full period.

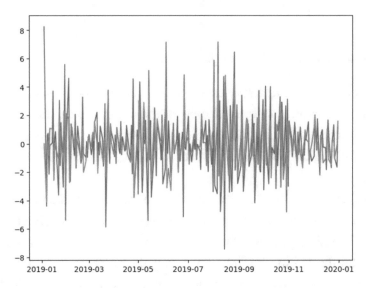

Figure 4-8. *Actuals vs. forecast on a train dataset*

This plot looks not too bad: the orange curve (predicted) follows the general trend in the blue curve (actual values). You can observe that the predicted curve is less extreme in its predictions (the highs are less high, and the lows are less low). The R2 score is **0.51**.

As an interpretation, you could observe that there is an underfit in this model: it does capture the basics, but it does not capture enough of the trend to be a very good model. At this stage, you can already have a check of its out-of-sample performance by creating a train and a test set. This can be done as follows using Listing 4-7.

Listing 4-7. Fitting the MA model on train data and evaluating the R2 score on train and test data

```
train = data.diff().dropna()[0:240]
test = data.diff().dropna()[240:250]

# Forecast the first MA(1) model
mod = ARIMA(train, order=(0,0,1))
res = mod.fit()

orig_data = data.diff().dropna()
pred = res.predict()
fcst = res.forecast(steps = len(test))

print(r2_score(train, pred))
print(r2_score(test, fcst))
```

This should give you a train R2 of **0.51** and a test R2 of **0.13**. This performance on the training data is not good, and the performance on the test data is even worse (only 13% of the variation in the test data can be explained by the MA model). A visual of the test data actuals vs. the forecast should confirm this bad fit (using Listing 4-8 to obtain Figure 4-9).

Listing 4-8. Plotting the out-of-sample forecast of the MA(1) model (MA with order 1)

```
plt.plot(list(test))
plt.plot(list(fcst))
plt.legend(['Actual Prices', 'Predicted Prices'])
plt.show()
```

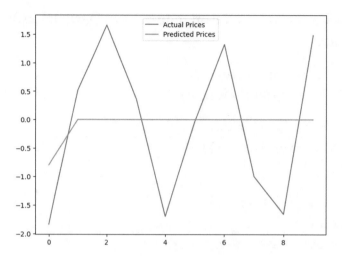

Figure 4-9. _Out-of-sample performance of the MA(1) model_

But when looking at this visual, you can see that there is something weird going on with the forecast on the test set. The MA(1) forecast has forecasted the average for every value except for the first step into the future.

And there is a mathematical reason for this: the MA(q) model uses the model error of the past q steps to predict the future. When predicting one step into the future, there is no problem: the model has the actual values and the fitted model of each time step before the prediction. But when doing a forecast for a second time step, the model does not know the actual values for time t + 1, while this is an input that is needed for forecasting t + 2.

Multistep Forecasting with Model Retraining

In the previous chapter, the forecast was made for ten steps forward. Yet no attention has yet been paid to an essential difference in forecasting. This is the difference between one-step forecasting and multistep forecasting.

One-step forecasting is the basis. It is relatively easy to predict one step forward. Multistep forecasting is not an easy task using time series, as errors will accumulate with every step forward you take.

In many cases, the most accurate solution is to retrain the model each time step. In fact, it is possible to do this if you update the model as soon as you obtain a new data point. This means that you would do repeated one-step forecasts. In some cases, this can work, while in other cases it is necessary to forecast further in the future.

It is important to consider the duration of your forecast in model evaluation. If you do multistep forecasts, you should do the evaluation on a multistep train-test split. If you do one-step forecasts, you should do the evaluation on a one-step train-test split.

If you do one-step forecasts, you will find yourself in a lack of a test dataset: you cannot compute a prediction error on a test data of one data point only. A solution that you can use is iteratively fitting the one-step model and constituting a multistep forecast based on multiple one-step forecasts. This is done in Listing 4-9.

Listing 4-9. Estimating the error of the MA(1) model for ten refitted one-step forecasts

```
import pandas as pd
train = data.diff().dropna()[0:240]
test = data.diff().dropna()[240:250]

# Import the ARMA module from statsmodels
from statsmodels.tsa.arima.model import ARIMA
fcst = []
for step in range(len(test)):
    # Forecast the first MA(1) model
    mod = ARIMA(train.reset_index(drop=True), order=(0,0,1))
    res = mod.fit()
    orig_data = data.diff().dropna()
    pred = res.predict()
    fcst += list(res.forecast(steps = 1))
    train = train.append(pd.Series(test[step]))
print(r2_score(list(test), fcst))
plt.plot(list(test))
plt.plot(fcst)
plt.legend(['Actual Prices', 'Predicted Prices'])
plt.show()
```

The R2 score that you obtain using this method is **0.48**: not perfect, but this starts to look like something useful. The graph of the forecasted values vs. the actual values is shown in Figure 4-10.

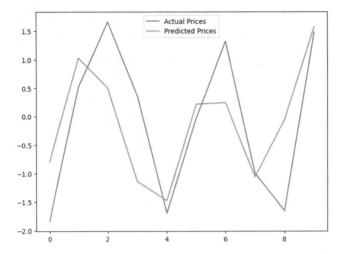

Figure 4-10. *Forecasting one step multiple times yields a relatively acceptable forecast*

Of course, the error estimate that you obtain this way is only the error estimate for the forecast of the next time step, and this will not be valid for a longer period. How to use this in practice depends on your specific use case.

For this specific example, you can consider that if you can perfectly predict the stock market of the next day, this is already quite the accomplishment: let's retain the one-step error in this case.

Grid Search to Find the Best MA Order

As a next step, let's try to do a grid search to find the order for the MA model that obtains the lowest error on the test set. Remember the grid search from the previous chapter: the basic version of the grid search simply consists of looping through each possible value of the order and evaluating the model on the test data.

There are other and better automated ways of optimizing hyperparameters: you will see them throughout the next chapters, so make sure that you understand the idea behind this basic version now. Now let's find out whether there is a specific order for the MA that delivers us better predictive performances using Listing 4-10.

Listing 4-10. Grid search to obtain the MA order that optimizes forecasting R2

```
def evaluate2(order):
    train = data.diff().dropna()[0:240]
    test = data.diff().dropna()[240:250]

    fcst = []
    for step in range(len(test)):
        # Forecast the first MA(1) model
        mod = ARIMA(train.reset_index(drop=True), order=(0,0,order))
        res = mod.fit()
        orig_data = data.diff().dropna()
        pred = res.predict()
        fcst += list(res.forecast(steps = 1))
        train = train.append(pd.Series(test[step]))

    return r2_score(list(test), fcst)

scores = []
for i in range(1, 21):
    scores.append((i, evaluate2(i)))

# observe best order is 4 with R2 of 0.566
scores = pd.DataFrame(scores)
print(scores[scores[1] == scores.max()[1]])
```

This will give you a best order of 4 with an out-of-sample R2 of **0.57**. You can have a look at the final forecast that this yields using the code in Listing 4-11. The plot that you will obtain is shown in Figure 4-11.

Listing 4-11. Obtaining the final forecast

```
train = data.diff().dropna()[0:240]
test = data.diff().dropna()[240:250]

fcst = []
for step in range(len(test)):
    # Forecast the first MA(1) model
    mod = ARIMA(train.reset_index(drop=True), order=(0,0,4))
    res = mod.fit()
```

```
    orig_data = data.diff().dropna()
    pred = res.predict()

    fcst += list(res.forecast(steps = 1))

    train = train.append(pd.Series(test[step]))

print(r2_score(list(test), fcst))

plt.plot(list(test))
plt.plot(fcst)
plt.legend(['Actual Prices', 'Forecasted Prices'])
plt.show()
```

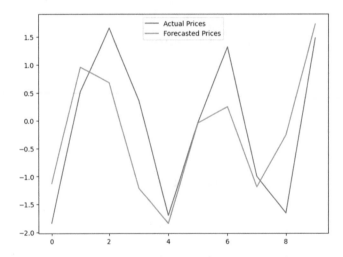

Figure 4-11. *The final result: the MA(4) forecast*

Key Takeaways

- The Moving Average model predicts the future based on impulses in the past.

    - Those impulses are measured as model errors.

    - The idea behind this is that unexpected impacts can actually have a large impact on the future.

- There is no one-shot computation for the MA model coefficients, so it takes a bit longer to estimate this model compared to the AR model.

- The MA model is the second building block of the SARIMA model. The AR and MA models together form the ARMA model, the topic of the next chapter.

- Multistep predictions are predictions of multiple time steps into the future. This is difficult with MA models, especially when the order is low. A solution can sometimes be to do multiple one-step predictions and retrain the model every time that you receive the new data.

- The autocorrelation function and the partial autocorrelation function can help you decide whether you are looking at an AR or an MA forecast. AR forecasts see their autocorrelation exponentially decay to 0 and alternate between positive and negative. MA autocorrelation functions are characterized by many values of almost zero and a few spikes.

CHAPTER 5

The ARMA Model

In this chapter, you will discover the ARMA model. It is a combination of the AR model and the MA model, which you have seen in the two previous chapters. Since the theory of this chapter should be already relatively familiar to you, this chapter will also introduce a few additional model quality indicators that are valid for the ARMA model but can also be used for the AR and MA models separately.

The Idea Behind the ARMA Model

The ARMA model is a simple combination of the AR and MA models. The idea behind combining the AR and MA models into one model is that the models together are more performant than one model. The development over time of one variable can follow multiple processes at the same time.

Let's do a short recap of the AR and MA models separately. The AR model tries to predict the future by **multiplying past values by a coefficient**. This AR process is based on the presence of autocorrelation: the target variable's present value is correlated to a past value. The MA model does not use past values of the target variable to predict its future, but rather uses the **error of past predictions** as an impulse for forecasted values.

The ARMA model is nothing more than a model that allows both these processes to be in place at the same time:

- Part of the future of a variable is explained by past values (the AR effect).

- Part of it is explained by past errors (the MA effect).

The fact that both are combined into one model makes it simply easier to use: you can imagine that working with two separate models would be cumbersome.

© Joos Korstanje 2021
J. Korstanje, *Advanced Forecasting with Python*, https://doi.org/10.1007/978-1-4842-7150-6_5

The Mathematical Definition of the ARMA Model

The ARMA model is defined by the following equation:

$$X_t = c + \varepsilon_t + \sum_{i=1}^{p} \varphi_i X_{t-i} + \sum_{i=1}^{q} \theta_i \varepsilon_{t-i}$$

In this equation, you clearly observe the AR part and the MA part. The AR part uses coefficients φ_i (phis) multiplied by past observations X_{t-i} for a **number of lags** defined as **p,** also called **the AR order**. The MA part uses coefficients θ_i (thetas) multiplied by past errors ε_{t-i} (epsilon) for a **number of lags q, the MA order**.

Those p and q are the same hyperparameters that you have encountered before. For model notation, you use *AR(p) for an AR model of order p* and *MA(q) for an MA model of order q*. In the ARMA model, p and q remain the same. This gives you the following notation: **ARMA(p, q).**

You should note here that there is no dependency between p and q, as there is no dependency between the AR and MA definitions in the equation. This means that an ARMA(p, q) model can take any order for p and q. Special cases are when p = 0, which means you have an MA model, and when q = 0, which means you have an AR model.

As p and q are hyperparameters, they are parameters that should be chosen by the modeler, while c, φ, and θ are coefficients of the model and will therefore be estimated by the model.

An Example: Predicting Sunspots Using ARMA

Now, let's get to work on an example and see whether the combined power of the AR and the MA model can deliver a good forecast. You will be working with the sunspot dataset, a relatively famous dataset that contains very long historical observations of the number of spots on the sun. You can get the data into Python using Listing 5-1 and see the head of the dataframe in Figure 5-1.

Listing 5-1. Getting the sunspot data into Python

```
import pandas as pd
data = pd.read_csv('Ch05_Sunspots_database.csv', usecols = [1, 2])
```

The sunspot dataset is well known for having a specific pattern of seasonal effects over 11-year periods. The dataset is not very detailed: it just contains the data counts per month. Since there is a lot of data to work with, let's set the goal to do a forecast with yearly data. To do this, the data need to be aggregated to yearly data. You can use Listing 5-2 to do this.

Listing 5-2. Aggregating the sunspot data to yearly data

```
data['year'] = data.Date.apply(lambda x: x[:4])
data = data[['Monthly Mean Total Sunspot Number',
'year']].groupby('year').sum()
data.head()
```

year	Monthly Mean Total Sunspot Number
1749	1618.5
1750	1668.0
1751	953.3
1752	956.0
1753	613.5

Figure 5-1. *An extract of the sunspot data*

Let's also make a plot over time to see what type of variation you are working with. You can do this using Listing 5-3. Using this code, you should obtain the graph in Figure 5-2.

Listing 5-3. Plotting the yearly sunspot data

```
import matplotlib.pyplot as plt
ax = data.plot()
ax.set_ylabel('Sunspots')
plt.show()
```

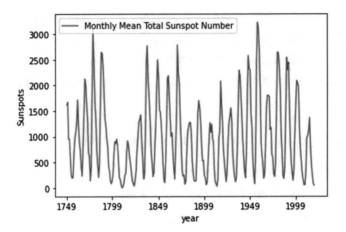

Figure 5-2. *A plot of the sunspot data*

Looking at this plot, the pattern is already strikingly clear: high peaks on regular intervals. Let's see how to use the ARMA model to capture this pattern and make predictions for the future. The first step, as always, is to verify whether the data is stationary or not.

Remember that stationarity means that there is no long-term trend in the data: the average is constant over time. When looking at the data, there is no obvious long-term trend, yet you can use the ADF test (Augmented Dickey Fuller test) to confirm this for you. Listing 5-4 shows you how you can do this.

Listing 5-4. Applying the ADF test to the sunspot yearly totals

```
from statsmodels.tsa.stattools import adfuller

result = adfuller(data['Monthly Mean Total Sunspot Number'])
print(result)

pvalue = result[1]

if pvalue < 0.05:
    print('stationary')
else:
    print('not stationary')
```

The result that you obtain when applying this test is unexpected: the ADF test tells you that the data is **not stationary**. In this case, the intuitive decision would be to assume stationarity, while the ADF tells you not to. A choice has to be made.

For now, let's keep it the easiest possible and stay with the original data.

Now the second thing to look at is the autocorrelation function (ACF) and the partial autocorrelation function (PACF). Remember that the ACF and PACF plots can help you to define whether you are working with AR or MA processes. You can obtain the ACF and PACF using Listing 5-5.

Listing 5-5. Creating the ACF and PACF plots

```
from statsmodels.graphics.tsaplots import plot_acf, plot_pacf
import matplotlib.pyplot as plt

plot_acf(data['Monthly Mean Total Sunspot Number'], lags=40)

plot_pacf(data['Monthly Mean Total Sunspot Number'], lags=40)

plt.show()
```

You will obtain the graphs in Figures 5-3 and 5-4.

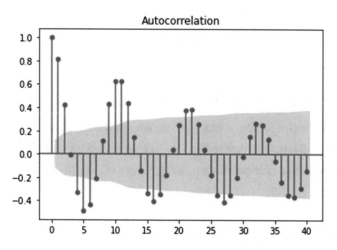

Figure 5-3. *The autocorrelation function (ACF)*

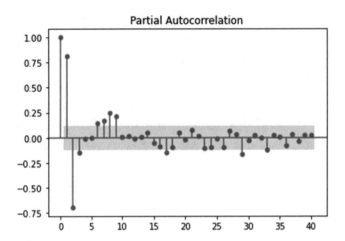

Figure 5-4. *The partial autocorrelation function (PACF)*

The ACF and PACF plots in Figures 5-3 and 5-4 are great cases to study time series patterns; it is rare to observe such strong patterns on real-life data. Remember from the previous chapters which type of patterns you should be looking at:

- AR processes are recognized by

  - Exponentially decaying partial autocorrelation

  - Partial autocorrelation decaying toward zero

  - Swings between negative and positive partial autocorrelation

- MA processes are identified by

  - Sudden spikes in the ACF and PACF

This means that in the sunspot case, you have a strong indicator for observing an AR pattern: there is exponential decay in the PACF, together with switching from negative to positive. There is no real evidence for an MA process when looking at those plots.

Fitting an ARMA(1,1) Model

The ARMA(p,q) model being simply the combination of the AR and MA models, it is important to realize that there is an optimization going on that finds the coefficients of the model that minimize the Mean Squared Error of the model. This is the same approach that has been discussed in Chapter 4.

Remember that there is an important difference between coefficients and hyperparameters. Coefficients are estimated by the model, whereas hyperparameters must be chosen by the modeler. In the ARMA(p,q) model, p and q are the hyperparameters that you must decide on using plots, performance metrics, and more.

To get started, let's see how to fit an ARMA(1,1) model in Python. An ARMA(1,1) model means an ARMA model with an AR component with order 1 and an MA component with order 1. Remember that the order refers to the number of historical values that are used to explain the current value.

For this first trial with order (1,1), the choice is just to start with the simplest ARMA model possible. And throughout the example, you'll see how to fine-tune this decision.

Applying an ARMA(1,1) model for the sunspot data means that you will explain tomorrow's value by looking at today's actual value (AR part) and today's error (MA part). You are not looking further back into the past, as the order is only 1.

This model can be created in Python using the code in Listing 5-6.

Listing 5-6. Fitting the ARMA(1,1) model

```python
from sklearn.metrics import r2_score
from statsmodels.tsa.arima.model import ARIMA

# Forecast the first ARMA(1,1) model
mod = ARIMA(list(data['Monthly Mean Total Sunspot Number']), order=(1,0,1))
res = mod.fit()
pred = res.predict()
print(r2_score(data, pred))

plt.plot(list(data['Monthly Mean Total Sunspot Number']))
plt.plot(pred)
plt.legend(['Actual Sunspots', 'Predicted Sunspots'])
plt.xlabel('Timesteps')
plt.show()
```

This code will also generate a plot of the actual values vs. the fitted values (Figure 5-5). It will also give you an R2 score of the model fit. Since you have not yet created a train-test split, this R2 score is not representative of any future performance. Yet it can be used to see how wrong the current model is: this gives an intuition on how to improve the model. In this case, with an R2 score of **0.76**, you can conclude that the model is already fitting not too badly.

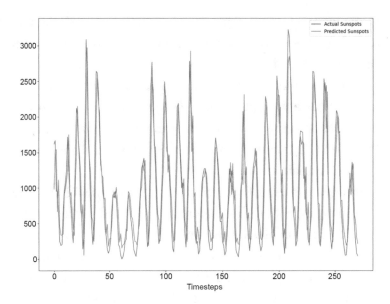

Figure 5-5. *The actual values vs. the fitted values of the ARMA(1,1)*

More Model Evaluation KPIs

As an addition to performance metrics, I want to take the opportunity to inspect a few more model fit KPIs. Until now, you have seen how to use the Augmented Dickey Fuller test to help you decide on differencing, and you have seen the ACF and PACF plots to help you decide on the type of model needed (AR vs. MA) and to help you choose the best order. You have also seen the train-test split combined with a hyperparameter search to identify the order that maximizes out-of-sample performance.

An additional KPI that can help you evaluate model fit is to look at the residuals of your model. If a model has a good fit, you will generally observe that the residuals follow a normal distribution. If you find the opposite, your model is generally missing out on important information.

You can make a histogram of the residuals of your model using the code in Listing 5-7.

Listing 5-7. Plotting a histogram of the residuals

```
ax = pd.Series(res.resid).hist()
ax.set_ylabel('Number of occurences')
ax.set_xlabel('Residual')
plt.show()
```

You should obtain the graph in Figure 5-6.

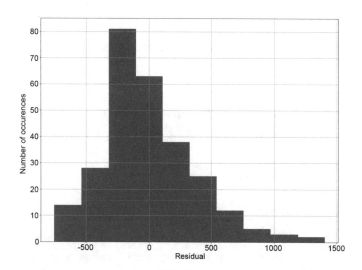

Figure 5-6. *Histogram of the residuals of the ARMA(1,1) model*

You can observe visually that the histogram of the residuals is not following the bell-shaped curve that is the **normal distribution**. This means that there are probably some lags to be added, which is logical: you already know that the sunspot data has an 11-year seasonality, and for now there is only one lagged period (1 year back) included in the model.

Note   There are many different approaches for testing the distribution of a variable. This includes histograms, QQ plots, and numerous hypothesis tests. Although hypothesis tests for normality are useful in many cases, the goal of a forecast is generally to maximize future performance. It is not necessary to have residuals that are perfectly normally distributed for this purpose, and the graphical way of identifying the normality of residuals is generally sufficient in practice.

When you've evaluated the normality of the residuals of your model, you can also look at the summary table of your model. You can obtain this summary table using the code in Listing 5-8.

Listing 5-8. Obtaining the summary table of your model's fit

```
res.summary()
```

You should obtain the summary table shown in Figure 5-7.

Dep. Variable:		y	No. Observations:		271
Model:	ARIMA(1, 0, 1)		Log Likelihood		-1987.170
Date:	Sun, 10 Jan 2021		AIC		3982.341
Time:	21:09:46		BIC		3996.749
Sample:	0		HQIC		3988.126
	- 271				
Covariance Type:	opg				

	coef	std err	z	P>|z|	[0.025	0.975]
const	984.7442	153.327	6.423	0.000	684.229	1285.259
ar.L1	0.7193	0.051	14.024	0.000	0.619	0.820
ma.L1	0.5254	0.053	9.922	0.000	0.422	0.629
sigma2	1.361e+05	1.09e+04	12.479	0.000	1.15e+05	1.58e+05

Ljung-Box (L1) (Q):	8.44	Jarque-Bera (JB):	41.94
Prob(Q):	0.00	Prob(JB):	0.00
Heteroskedasticity (H):	1.43	Skew:	0.81
Prob(H) (two-sided):	0.09	Kurtosis:	4.05

Figure 5-7. *Summary table of the ARMA(1,1) model*

Important pieces of information that you can get from this table are **the estimates of the coefficients** and the corresponding **p-values**. Remember that you are working with an ARMA(1,1) model and that you therefore will have one coefficient for the AR model (for the data point one time step back) and one coefficient for the MA model (for the error of one time step back).

When you fit the model in the data, your model is estimating the best possible values for those coefficients. You can see the estimates of the coefficients in the table on the line that says **ar.L1** (AR coefficient for the first lag is **0.7193**) and on the line that says **ma.L1** (MA coefficient for the first lag is **0.5254**).

You can interpret those by looking back to the formula. Basically, those estimates state that you can predict a future data point, by filling in the ARMA formula using the **const** as c, the ar.L1 as coefficient φ_i (phi), and the ma.L1 as coefficient θ_i (tetha).

Filling in those coefficients would **allow you to compute the next value**. This is great information for understanding what information is used by your model to predict the future.

Now that you know the estimated values of your coefficients, you can also look at the hypothesis tests for those coefficients. For each coefficient, the summary table shows a hypothesis test that tells you whether the coefficient is **significantly different from zero**. If a coefficient were to be zero, this would mean that it is not useful to include the parameter. Yet, if a hypothesis test proves that a coefficient is significantly different from zero, this means that it is useful to add the coefficient in your model. This helps you in deciding which order to use for your ARMA forecast.

The p-values of the hypothesis tests can be found on the line of the coefficient, under the column **P>|z|**. This column gives you the p-value of the hypothesis test. The p-value must be **smaller than 0.05** to prove that the coefficient is significantly different from zero.

In this case, the interpretation is as follows:

- The p-value for the ar.L1 coefficient is 0.000, so significant.

- The p-value for the ma.L1 coefficient is also 0.000, so significant.

- Both are smaller than 0.05, and therefore you can conclude that both coefficients should be retained in the model.

As you can imagine, this is very useful information when creating a model. You can go back and forward between **increasing and decreasing the order of AR and MA independently** and try to find at which point the coefficients start to become insignificant (p-values higher than 0.05) in which case you take the order of the highest lag that was still significant as the best model. Of course, keep in mind that significance is one indicator between multiple indicators.

Automated Hyperparameter Tuning

Although this theoretical approach can be very interesting, you should combine it with tests of predictive performance, as you have done already in Chapters 3 and 4 using very basic approaches to hyperparameter optimization.

If you remember, what you did was creating a train-test split and then looping through all possible values of the hyperparameter and evaluating the performance on the test set. The order that gave the best performance on the test set was retained.

In this part, you will make an addition to this approach by officializing the notions of grid search and cross-validation.

Grid Search: Tuning for Predictive Performance

I've purposely waited explaining **grid search** until here because it is better explained with an example with two hyperparameters. In the ARMA model, two hyperparameters need to be optimized at the same time: p and q.

The goal of a grid search for hyperparameter optimization is to make the task of choosing hyperparameters. You have seen multiple statistical tools that can help you to make the choice:

- Autocorrelation function (ACF)

- Partial autocorrelation function (PACF)

- Model residuals

- Summary table

With multiple indicators, it is difficult to decide on the best parameters. A grid search is a method that will test every combination of **hyperparameters** and evaluate the predictive error of this combination of hyperparameters. This will give you an objective estimate of the hyperparameters that will obtain the best performance.

A basic approach for a grid search is to make a **train-test split** and to fit a model with each combination of hyperparameters on the train set and evaluate the model on the test set. In practice, a grid search is often combined with **cross-validation**. As you've seen in Chapter 2, cross-validation is an augmentation of the train-test split approach in which the train-test split is repeated multiple times. This yields a more reliable error estimate.

To apply a grid search with cross-validation, you need to implement a loop through each combination of hyperparameters (in this case, p and q). For each combination, the code will split the data in multiple **cross-validation splits**: multiple combinations of train-test splits. Those splits are also called folds. For each of those folds, you train the model on the train set and test the model on the test set. This yields an error for each fold, of which you take the average to obtain an error for each hyperparameter combination. The hyperparameter combination with the best error will be the model that you retain for your forecast.

Listing 5-9 shows how this is done in code.

Listing 5-9. Grid search with cross-validation for optimal p and q

```python
import numpy as np
from sklearn.model_selection import TimeSeriesSplit
data_array = data.values

avg_errors = []

for p in range(13):
    for q in range(13):

        errors = []

        tscv = TimeSeriesSplit(test_size=10)

        for train_index, test_index in tscv.split(data_array):

            X_train, X_test = data_array[train_index], data_array[test_index]
            X_test_orig = X_test

            fcst = []
            for step in range(10):

                try:
                    mod = ARIMA(X_train, order=(p,0,q))
                    res = mod.fit()

                    fcst.append(res.forecast(steps=1))

                except:
                    print('errorred')
                    fcst.append(-9999999.)

                X_train = np.concatenate((X_train, X_test[0:1,:]))
                X_test = X_test[1:]
```

```
        errors.append(r2_score(X_test_orig, fcst))

    pq_result = [p, q, np.mean(errors)]

    print(pq_result)
    avg_errors.append(pq_result)

avg_errors = pd.DataFrame(avg_errors)
avg_errors.columns = ['p', 'q', 'error']
result = avg_errors.pivot(index='p', columns='q')
```

The result of the grid search is a table with an estimated error for each combination of p and q in the limits that you have specified (between 0 and 11). Thanks to the cross-validation approach, you can be confident in this error estimate: after all, it is the average of five separate error measurements, and this makes it unlikely that any bias has been caused by a favorable test set selection. You can see the results table in Figure 5-8.

q	0	1	2	3	4	5	6	7	8	9	10	11	12
p													
0	-0.166	0.528	6.670000e-01	7.010000e-01	7.390000e-01	0.747	0.747	0.741	7.550000e-01	7.690000e-01		0.717	0.767 0.777
1	0.608	0.704	7.150000e-01	7.330000e-01	7.280000e-01	0.745	0.757	0.747	7.490000e-01	7.670000e-01		0.775	0.786 0.770
2	0.770	0.784	7.850000e-01	7.860000e-01	7.890000e-01	0.790	0.808	0.812	8.030000e-01	8.260000e-01		0.821	0.818 0.816
3	0.785	0.779	7.800000e-01	8.250000e-01	8.200000e-01	0.817	0.816	0.815	8.070000e-01	8.200000e-01		0.815	0.818 0.810
4	0.784	0.779	7.790000e-01	7.830000e-01	8.160000e-01	0.816	0.815	0.811	8.110000e-01	8.160000e-01		0.817	0.823 0.817
5	0.780	0.776	7.820000e-01	7.930000e-01	8.170000e-01	0.830	0.820	0.822	8.220000e-01	8.250000e-01		0.822	0.820 0.829
6	0.779	-3207013.667	-1.002725e+07	-5.488856e+06	7.970000e-01	0.826	0.821	0.822	8.190000e-01	8.210000e-01		0.829	0.828 0.815
7	0.785	0.805	-3.206515e+06	-1.370380e+07	-1.552040e+07	0.828	0.830	0.829	8.270000e-01	8.310000e-01		0.821	0.831 0.815
8	0.794	0.809	8.110000e-01	-6.322078e+06	7.900000e-01	0.814	0.836	0.829	8.130000e-01	8.380000e-01		0.825	0.821 0.817
9	0.819	0.818	8.170000e-01	8.170000e-01	7.940000e-01	0.821	0.813	0.820	8.100000e-01	8.430000e-01		0.821	0.821 0.831
10	0.817	0.817	8.180000e-01	8.150000e-01	8.010000e-01	0.807	0.828	0.833	8.100000e-01	8.480000e-01		0.836	0.805 0.789
11	0.817	0.816	8.150000e-01	8.100000e-01	8.100000e-01	0.798	0.797	0.812	-1.002761e+07	-4.150568e+07	-3736028.355	0.814	0.785
12	0.815	0.814	8.140000e-01	8.230000e-01	8.000000e-01	0.823	-314.510	0.818	8.150000e-01	8.430000e-01		0.815	0.798 0.787

Figure 5-8. *Table with the R2 scores for each combination of p and q*

The combination that has yielded the best R2 score is the combination of **p = 10 and q = 9**, which yields an estimated R2 score on the test set of **0.84**. As a final step, you can refit the model on a train-test split to plot this on a train-test set to see what this error looks like. This is done in Listing 5-10 and should give you the graph displayed in Figure 5-9.

Listing 5-10. Showing the test prediction of the final model

```
data_array = data.values
X_train, X_test = data_array[:-10], data_array[-10:]
X_test_orig = X_test

fcst = []
for step in range(10):
    mod = ARIMA(X_train, order=(10,0,9))
    res = mod.fit()
    fcst.append(res.forecast(steps=1))
    X_train = np.concatenate((X_train, X_test[0:1,:]))
    X_test = X_test[1:]

plt.plot(X_test_orig)
plt.plot(fcst)
plt.legend(['Actual Sunspots', 'Predicted Sunspots'])
plt.xlabel('Time steps of test data')
plt.show()
```

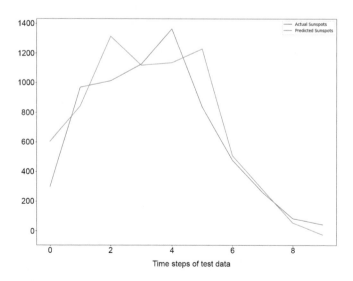

Figure 5-9. *Prediction on the test set by the final model*

Key Takeaways

- The ARMA model is the combination of the autoregressive model and the Moving Average model.

- The hyperparameters of ARMA are p and q: p for the autoregressive order and q for the moving average order.

- A grid search is a common tool for optimizing the choice of hyperparameters. It can be combined with cross-validation to yield more reliable error estimates.

- You can use the distribution of residuals to evaluate the fit of a model. If the residuals do not follow a normal distribution, there is generally something wrong with the model specification.

- You can use the model summary to look at detailed indicators of model fit. You can also find the estimates of model coefficients and a hypothesis test that tests the model coefficients against zero.

CHAPTER 6

The ARIMA Model

Having seen several building blocks of univariate time series, in this chapter, you're going to see a model that combines even more of the time series components: the ARIMA model.

To keep track of the different building blocks, let's get back to the table of time series components (Table 6-1) to see where the ARIMA model is at.

Table 6-1. *The Building Blocks of Univariate Time Series*

Name	Explanation	Chapter
AR	Autoregression	3
MA	Moving Average	4
ARMA	Combination of AR and MA models	5
ARIMA	Adding differencing (I) to the ARMA model	6
SARIMA	Adding seasonality (S) to the ARIMA model	7
SARIMAX	Adding external variables (X) to the SARIMA model *(note that external variables make the model not univariate anymore)*	8

As you can see, ARIMA is getting close to the final model. The only parts that are not yet included are S (seasonality) and X (external variables). Some time series processes do not have seasonality nor external variables, which makes ARIMA frequently used as a stand-alone model.

In practice, if you are doing univariate time series, you will generally go directly to ARIMA if you have no expectation of seasonality, or if you expect seasonality, you will go directly to SARIMA (which will be covered in the next chapter).

© Joos Korstanje 2021
J. Korstanje, *Advanced Forecasting with Python*, https://doi.org/10.1007/978-1-4842-7150-6_6

ARIMA Model Definition

The ARIMA model is a model that combines the AR and MA building blocks, just as you have seen with ARMA in the previous chapter. The addition in ARIMA is the "I" building block, which stands for **automatic differencing of non-stationary time series**.

You should remember that during the past chapters, stationarity has been presented as an important concept in time series. A stationary time series is a time series that has no long-term trend. If a time series is not stationary, you can make it stationary by applying differencing: replacing the actual values by the difference between the actual and the previous value.

The "I" in ARIMA is for **integrating**. Integrating is a more mathematical synonym for differencing a non-stationary time series. In ARIMA, this differencing is not anymore done in advance of the modeling phase, but it is done during the model fit.

Model Definition

The ARIMA model is short for the Autoregressive, Integrated Moving Average model. The difference from the ARMA model is small: there is just an additional effect that makes the time series non-stationary. A simple example of this would be a linear increasing trend, as shown in the following equation:

$$X_t = c + \varepsilon_t + \sum_{i=1}^{p} \varphi_i X_{t-i} + \sum_{i=1}^{q} \theta_i \varepsilon_{t-i} + \delta t$$

When differencing a time series, you actually start to model the differences from one step to another rather than the original values. If the actual values of a variable are not stable over time, it is still possible that the differences are stable over time.

The linear trend is a great example of this. Imagine a linear trend that starts from 0 and increments by 2 every time step. The values will not be stationary at all: they will augment infinitely. Yet the difference between each value and the next is always 2, so the differenced time series is perfectly stationary.

ARIMA on the CO2 Example

The fact that the differencing is part of the hyperparameters has a great added value for model building. This makes it possible to do automated hyperparameter tuning on the number of times that differencing should be applied. Let's see this with an example.

The data in this example are weekly CO2 data that are available through the statsmodels library. You can obtain the data using the code in Listing 6-1.

Listing 6-1. Importing the data

```
import statsmodels.api as sm
data = sm.datasets.co2.load_pandas()
data = data.data
data.head()
```

To get an idea of the data that you're working with, as always, it is good to make a plot of the data over time. Use the code in Listing 6-2 to obtain the graph in Figure 6-1.

Listing 6-2. Plotting the data

```
import matplotlib.pyplot as plt
ax = data.plot()
ax.set_ylabel('CO2 level')
plt.show()
```

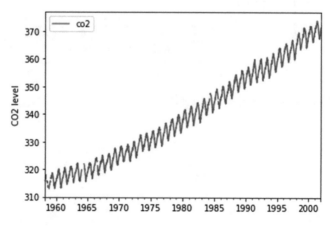

Figure 6-1. *Plot of the CO2 data over time*

This data shows a very obvious sign of an upward trend, which is fairly constant. And there is also a very clear seasonality pattern in this data (up and down).

Now remember, from the previous chapters, the first thing that you should generally look at is autocorrelation and partial autocorrelation functions. Remember that the autocorrelation (ACF) and partial autocorrelation (PACF) plots are relevant only when applied to stationary data. You can use Listing 6-3 to create ACF and PACF plots directly on the differenced data.

Listing 6-3. ACF and PACF plots

```
from statsmodels.graphics.tsaplots import plot_acf, plot_pacf
plot_acf(data.diff().dropna(), lags=40)
plot_pacf(data.diff().dropna(), lags=40)
plt.show()
```

You should obtain the ACF plot shown in Figure 6-2 and the PACF plot in Figure 6-3.

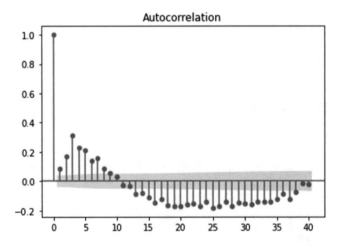

Figure 6-2. *Autocorrelation function plot*

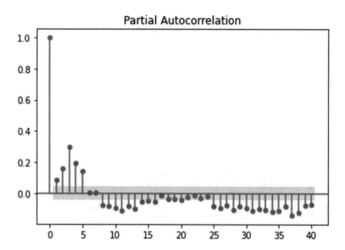

Figure 6-3. *Partial autocorrelation function plot*

It is interesting to note that there are many lagged values in the autocorrelation and partial autocorrelation. So what does this mean? In general, if there is no decay of the correlations toward zero, this means the data is not stationary. Yet the data has been differenced and seems stationary.

The answer here is that the decay occurs relatively late in the autocorrelation. The ACF and PACF plots have 40 lags, but that is not enough for the current example. You can use Listing 6-4 to obtain ACF and PACF plots that go further back, and you will observe that there is a decay toward zero. The respective plots are shown in Figures 6-4 and 6-5.

Listing 6-4. ACF and PACF plots with more lags

```
plot_acf(data.diff().dropna(), lags=600)
plot_pacf(data.diff().dropna(), lags=600)
plt.show()
```

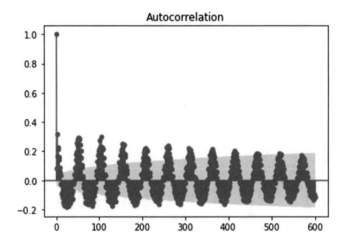

Figure 6-4. *Autocorrelation function plot with 600 lags*

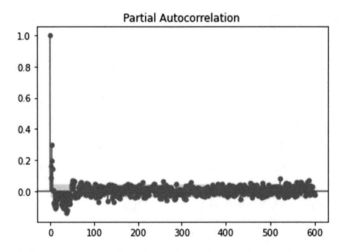

Figure 6-5. *Partial autocorrelation function plot with 600 lags*

This makes it an interesting case for the question of the order of the model. Let's see how to augment the previous chapter's grid search cross-validation by adding the I as a third hyperparameter to use in the optimization. The code for this can be seen in Listing 6-5.

Listing 6-5. Hyperparameter tuning

```
import pandas as pd
from statsmodels.tsa.arima.model import ARIMA
from sklearn.metrics import r2_score
```

```python
import numpy as np
from sklearn.model_selection import TimeSeriesSplit
data_array = data[['co2']].values

avg_errors = []

for p in range(6):
    for q in range(6):
        for i in range(3):
            errors = []

            tscv = TimeSeriesSplit(test_size=10)

            for train_index, test_index in tscv.split(data_array):

                X_train, X_test = data_array[train_index],
                data_array[test_index]
                X_test_orig = X_test

                fcst = []
                for step in range(10):

                    try:
                        mod = ARIMA(X_train, order=(p,i,q))
                        res = mod.fit()

                        fcst.append(res.forecast(steps=1))

                    except:
                        print('errorred')
                        fcst.append(-9999999.)

                    X_train = np.concatenate((X_train, X_test[0:1,:]))
                    X_test = X_test[1:]

                errors.append(r2_score(X_test_orig, fcst))

            pq_result = [p, i, q, np.mean(errors)]

            print(pq_result)
            avg_errors.append(pq_result)
```

```
avg_errors = pd.DataFrame(avg_errors)
avg_errors.columns = ['p', 'i', 'q', 'error']
avg_errors.sort_values('error', ascending=False)
```

The result of this code will give you a dataframe ordered by R2 scores. The best R2 score is **0.741**, and it is obtained for the combination of **p = 4, q = 4,** and **I = 1**. This means that there is an autoregressive effect of order 4 and a moving average effect of order 4. The data must be differenced one time. Listing 6-6 shows you how to show the test fit of the model, and the graph is shown in Figure 6-6.

Listing 6-6. Plot the final result

```
X_train, X_test = data_array[:-10], data_array[-10:]
X_test_orig = X_test

fcst = []
for step in range(10):

    mod = ARIMA(X_train, order=(4,1,4))
    res = mod.fit()
    fcst.append(res.forecast(steps=1))
    X_train = np.concatenate((X_train, X_test[0:1,:]))
    X_test = X_test[1:]

plt.plot(fcst)
plt.plot(X_test_orig)
plt.legend(['Predicted', 'Actual'])
plt.ylabel('CO2 Level')
plt.xlabel('Time Step of Test Data')
plt.show()
```

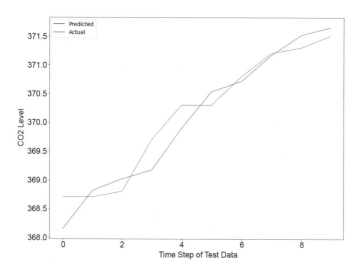

Figure 6-6. *Result of the best forecast*

Key Takeaways

- The ARIMA model combines three effects:

  - The AR process, based on autocorrelations between past and present values

  - The MA process, based on correlations between past errors and present values

  - Automatic integration of a time series if it is not stationary

- The ARIMA(p,I,q) model has three hyperparameters:

  - The order of the AR process denoted by p

  - The order of the MA process denoted by q

  - The order of integration denoted by I (or d in some notations)

CHAPTER 7

The SARIMA Model

In this chapter, you are going to discover the final model of univariate time series models: the SARIMA model. This will be the chapter in which everything on univariate time series comes together.

Univariate Time Series Model Breakdown

With the ARIMA components covered in previous chapters, you can now model autoregressive processes, moving average components, and integration (differencing). A common process of time series is still missing from this: **seasonality**.

Let's have a look at the univariate time series breakdown table to see where this fits in (Table 7-1). You can see that there is only one model more complicated than the SARIMA model: the SARIMAX model.

Table 7-1. *The Building Blocks of Univariate Time Series*

Name	Explanation	Chapter
AR	Autoregression	3
MA	Moving Average	4
ARMA	Combination of AR and MA models	5
ARIMA	Adding differencing (I) to the ARMA model	6
SARIMA	Adding seasonality (S) to the ARIMA model	7
SARIMAX	Adding external variables (X) to the SARIMA model *(note that external variables make the model not univariate anymore)*	8

© Joos Korstanje 2021
J. Korstanje, *Advanced Forecasting with Python*, https://doi.org/10.1007/978-1-4842-7150-6_7

The SARIMAX model is quite different from SARIMA: it uses external variables that correlate with the target variable. This makes SARIMAX a very powerful tool as well, but it can be applied only if you have external variables. Using external variables makes the model not a univariate time series model.

The power of the SARIMA model is that it needs to use only the history of the target variable, which can be applied to very many cases of forecasting. Within univariate time series modeling, the SARIMA model is the most complete model, using AR, MA, integration for modeling trends, and seasonality.

The SARIMA Model Definition

The SARIMA model combines many different processes in one and the same model. Therefore, the model definition is going to be relatively complex mathematically. The following equation is the definition of the SARIMA model:

$$y_t = u_t + \eta_t$$
$$\varphi_p(L)\tilde{\phi}_p(L^s)\Delta^d \Delta_s^D u_t = A(t) + \theta_q(L)\tilde{\theta}_Q(L^s)\zeta_t$$

In this model, you can observe the following parts:

- The regular AR part, with the coefficients φ_p (small phis)
- The seasonal AR part, with the coefficients $\tilde{\phi}_P$ (capital phis)
- The regular MA part, with the coefficients θ_q (small thetas)
- The seasonal MA part, with the coefficients $\tilde{\theta}_Q$ (capital thetas)
- The regular integration part, indicated by order d
- The seasonal integration part, indicated by order D
- Coefficient of seasonality s

As you can see, adding seasonality to the ARIMA model makes the model much more complex. The seasonality is not just one building block added to the previous model. There is a seasonal part added for each of the building blocks. This makes the lag selection of the SARIMA model also much more difficult. The same hyperparameters as before still need to be decided on:

- p for the AR order

- q for the MA order

- I for the differencing order

To this are added three more orders:

- P for the seasonal AR order

- Q for the seasonal MA order

- D for the seasonal differencing

And lastly, the choice for parameter **s, the seasonal period**, must be made. This s is not a hyperparameter like the other orders. s should not be tested using a grid search; rather, it should be decided based on logic. The s is the periodicity of the seasonality. If you work with monthly data, you should choose 12; if you work with weekly data, you should choose 52; etc.

Example: SARIMA on Walmart Sales

In this chapter, you will apply the SARIMA model to a Walmart sales forecasting dataset that is available on Kaggle (`www.kaggle.com/c/walmart-recruiting-store-sales-forecasting/`).

Let's get this data in Python and start to see what it looks like using Listing 7-1. Note that the Walmart data contains many stores and products, but for this exercise, you will take the sum of the weekly data.

Listing 7-1. Importing the data and creating a plot

```
import pandas as pd
import matplotlib.pyplot as plt
data = pd.read_csv('train.csv')
data = data.groupby('Date').sum()
ax = data['Weekly_Sales'].plot()
ax.set_ylabel('Weekly sales')
plt.gcf().autofmt_xdate()
plt.show()
```

This will give you the plot shown in Figure 7-1.

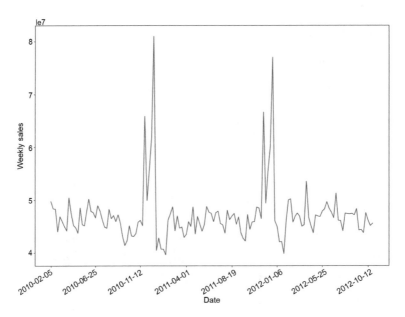

Figure 7-1. *Walmart sales over time*

Let's first try a first model: SARIMA(1,1,1)(1,1,1)52. This notation indicates an order 1 for each of the hyperparameters. The seasonality is 52, because you're working with weekly data. You will be forecasting ten steps out rather than doing ten one-step forecasts. This is a case of multistep forecasting. Listing 7-2 shows how to fit the model and obtain a plot of the performances on the test set (Figure 7-2). You will also obtain the R2 score on the test data.

Listing 7-2. Fitting a SARIMA(1,1,1)(1,1,1)52 model

```
import random
random.seed(12345)
import statsmodels.api as sm
from sklearn.metrics import r2_score

train = data['Weekly_Sales'][:-10]
test = data['Weekly_Sales'][-10:]
mod = sm.tsa.statespace.SARIMAX(data['Weekly_Sales'][:-10], order=(1,1,1),
seasonal_order=(1,1,1,52))
```

```
res = mod.fit(disp=False)
fcst = res.forecast(steps=10)

plt.plot(list(test))
plt.plot(list(fcst))
plt.legend(['Actual data', 'Forecast'])
plt.ylabel('Sales')
plt.xlabel('Test Data Time Step')
plt.show()
r2_score(test, fcst)
```

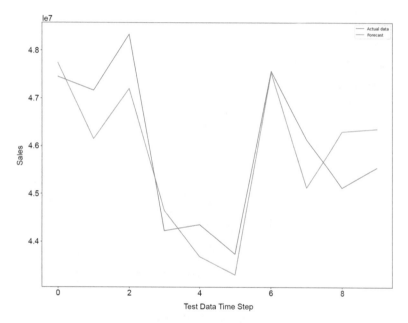

Figure 7-2. *Predictive performance of the SARIMA model*

In this version, without any hyperparameter optimization, the SARIMA model obtains an R2 score of **0.73**. This means that the model can explain 73% of the variation in the test data. This is a very promising score.

The reason that this score is already quite good is probably the flexibility of the SARIMA model. The fact that it is a combination of multiple effects makes it fit to many different underlying processes. You will now see how this score can improve using a grid search of the hyperparameters.

To speed up the code, you will not use cross-validation in the grid search here. Speed of execution is also an important aspect of machine learning. As a second optimization, you will be working with a ten-step forecast rather than with ten updated steps of a one-step forecast. In this way, it will be interesting to see how the seasonal effects may be more efficient in capturing long-term trends. This will be done in Listing 7-3.

Note that a try-except block has been added to avoid the code stopping in case an error occurs while fitting the model. Depending on your hardware, some of the fits may be too heavy and cause out-of-memory errors. This is not a problem, as it will still allow you to find the best possible combination of hyperparameters.

Listing 7-3. Grid Search on the SARIMA model

```
scores = []
for p in range(2):
    for i in range(2):
        for q in range(2):
            for P in range(2):
                for D in range(2):
                    for Q in range(2):

                        try:
                            mod = sm.tsa.statespace.SARIMAX(train,
                                order=(p,0,q), seasonal_order=(P,D,Q,52))
                            res = mod.fit(disp=False)

                            score = [p,i,q,P,D,Q,r2_score(test, res.
                                forecast(steps=10))]
                            print(score)
                            scores.append(score)

                            del mod
                            del res

                        except:
                            print('errored')

res = pd.DataFrame(scores)
res.columns = ['p', 'i', 'q', 'P', 'D', 'Q', 'score']
res.sort_values('score')
```

You see that the best order here is $(0,1,1)(1,1,1)$. The zero means that there is no regular AR process. There is a seasonal AR process. A regular and a seasonal MA process are present. Regular and seasonal integration is needed.

This model gives you an R2 on the test data of **0.734**, just slightly better than the nonoptimized model. As the last step, you can check out the plot of this forecast on the test data. This can be done by updating the order in Listing 7-2, and this will allow you to obtain the plot in Figure 7-3.

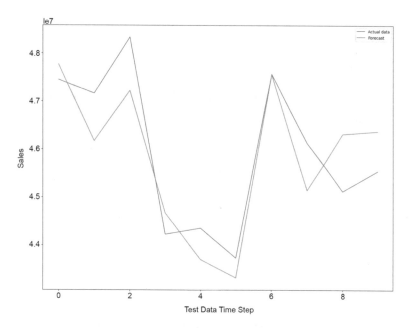

Figure 7-3. *Predictive performance of the optimized SARIMA model*

The grid search that was presented here is a good approach to tuning a SARIMA model. At this point, I want to point to a function called **auto_arima** from the **pyramid** library that automatically tunes ARIMA and SARIMA models. Using this function, you have less code to write for the same optimization. You can check out how to use auto_ arima over here: https://pypi.org/project/pyramid-arima/.

Key Takeaways

- The SARIMA model adds a seasonal effect to the ARIMA model.

- The SARIMA model is the completest version of univariate time series models.

- There are four more hyperparameters in the SARIMA model:

  - Seasonal AR order

  - Seasonal MA order

  - Seasonal integration order

  - Periodicity (based on the number of periods that a seasonality would logically return)

PART III

Multivariate Time Series Models

CHAPTER 8

The SARIMAX Model

In this chapter, you will discover the SARIMAX model. This model is the most complete version of classical time series models, as it contains all of the components that you've discovered throughout the previous chapters of this book. It adds the X component: external variables.

Time Series Building Blocks

Let's have a quick look back at the different components of time series that you have seen throughout the previous chapters using Table 8-1.

Table 8-1. *The Building Blocks of Univariate Time Series*

Name	Explanation	Chapter
AR	Autoregression	3
MA	Moving Average	4
ARMA	Combination of AR and MA models	5
ARIMA	Adding differencing (I) to the ARMA model	6
SARIMA	Adding seasonality (S) to the ARIMA model	7
SARIMAX	Adding external variables (X) to the SARIMA model *(note that external variables make the model not univariate anymore)*	8

The chapters build up time series models from easy to complex. The SARIMAX model is not considered a perfect example of a univariate time series model. Univariate time series models only use variation in the target variable, while the SARIMAX model uses external variables as well.

J. Korstanje, *Advanced Forecasting with Python*, https://doi.org/10.1007/978-1-4842-7150-6_8

Model Definition

The mathematical definition of the SARIMAX model is as follows:

$$y_t = \beta_t x_t + u_t$$
$$\varphi_p(L)\tilde{\phi}_p(L^s)\Delta^d\Delta^D_s u_t = A(t) + \theta_q(L)\tilde{\theta}_Q(L^s)\zeta_t$$

The β (beta) part in the first formula represents the external variable. For the rest, the model is very similar to the SARIMA model, but for completeness, let's relist the hyperparameters of this model:

- p for the AR order

- q for the MA order

- I for the differencing order

- P for the seasonal AR order

- Q for the seasonal MA order

- D for the seasonal differencing

- s for the seasonal coefficients

Supervised Models vs. SARIMAX

In further chapters, you'll see many cases of supervised models. As you may recall from the first chapter, supervised models use external variables to predict a target variable. This idea is very similar to the X part in SARIMAX.

The question, of course, is which one of them you should use in practice. From a theoretical point of view, SARIMAX may be preferred in cases where the time series part is more present than the external variables part. If the external variables alone can explain a lot and this is complemented by a part of autocorrelation or seasonality, supervised models may be the better choice.

Yet, as always, the reasonable thing to do is to use multiple models in a model benchmark. The choice for a model should then simply be based on the predictive performance of the model.

Example of SARIMAX on the Walmart Dataset

In the previous chapter, you've seen the SARIMA model used to predict weekly sales at Walmart. Yet the dataset does not just contain weekly sales data: there is also an indicator that tells you whether a week did or did not have a holiday in it.

You did not use the holiday information in the SARIMA model, as it is impossible to add external data to it. And this did not really matter as there were no extreme peaks due to holidays in the test data. Even though the holiday information was missing in the model, it was not represented in the error estimate.

Yet it is important to add such information, as it may be able to explain a large part of the variation in the model. Let's see how to improve on the Walmart example using external variables.

As a first step, you should import and prepare the data. You can use Listing 8-1 for this. This code also shows you how to create a plot of sales over time.

Listing 8-1. Preparing the data and making a plot

```python
import pandas as pd
import matplotlib.pyplot as plt
data = pd.read_csv('walmart/train.csv')
data = data.groupby('Date').sum()
data['IsHoliday'] = data['IsHoliday'] > 0
data['IsHoliday'] = data['IsHoliday'].apply(
    lambda x: float(x)
)

ax = data['Weekly_Sales'].plot()
ax.set_ylabel('Weekly Sales')
plt.gcf().autofmt_xdate()
plt.show()
```

You should obtain the graph displayed in Figure 8-1.

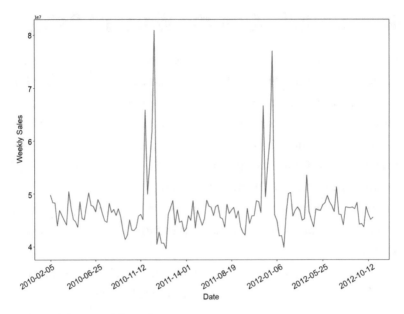

Figure 8-1. *Plot of the weekly sales*

The next step is to use correlation analysis to study whether we may expect an improvement from adding the holiday information into the model. If there is no correlation at all between sales and holidays, it would be unwise to add it to the model. Let's use Listing 8-2 to compute the correlation coefficient.

Listing 8-2. Is there a correlation between sales and holidays?

```
data[['Weekly_Sales', 'IsHoliday']].corr()
```

This will output a correlation matrix as shown in Figure 8-2.

	Weekly_Sales	IsHoliday
Weekly_Sales	1.000000	0.172683
IsHoliday	0.172683	1.000000

Figure 8-2. *The correlation matrix*

The correlation coefficient between sales and holidays is **0.17**. This is not very high, but enough to consider adding the variable to the model. The model itself will compute the most appropriate coefficient for the variable.

To create a SARIMAX model, you can use Listing 8-3. It is important to note the terminology of **endog** and **exog**:

- **Endogenous variables** (endog) are the target variable. This is where all the time series components will be estimated from. In the current case, it is the weekly sales.

- **Exogenous variables** (exog) are explanatory variables. This is where the model takes additional correlation from. In the current example, this is the holiday variable.

For this particular example, I will not do the hyperparameter search again. If you remember from the previous chapter, the optimal score was found using SARIMA(0,1,1) (1,1,1),52. So for this example, let's use SARIMAX(0,1,1)(1,1,1)52. The code also computes an R2 score and will show the plot of performance on a test set of 10 weeks.

Listing 8-3. Fitting a SARIMAX model

```
import random
random.seed(12345)
import statsmodels.api as sm
from sklearn.metrics import r2_score

train = data['Weekly_Sales'][:-10]
test = data['Weekly_Sales'][-10:]

mod = sm.tsa.statespace.SARIMAX(
    endog=data['Weekly_Sales'][:-10],
    exog=data['IsHoliday'][:-10],
    order=(0,1,1),
    seasonal_order=(1,1,1,52),
)
res = mod.fit(disp=False)
fcst = res.forecast(steps=10, exog = data['IsHoliday'][-10:])

plt.plot(list(test))
plt.plot(list(fcst))
plt.xlabel('Steps of the test data')
plt.ylabel('Weekly Sales')
```

```
plt.legend(['test', 'forecast'])
plt.show()
r2_score(test, fcst)
```

The R2 score obtained by this forecast is **0.734**. The performance can be seen visually in Figure 8-3.

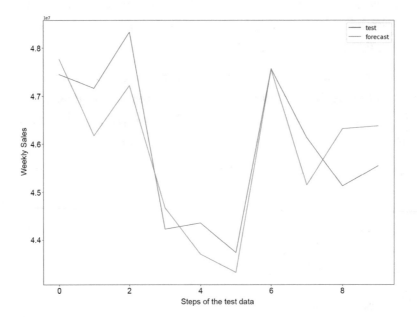

Figure 8-3. *Predictive performance on the test data*

This chapter has shown you how to use the SARIMAX model using one additional variable. Yet remember that there is no limit for the number of exogenous variables included in a SARIMAX model. As long as you can provide future values for this variable, this may work.

Pay attention here: many variables that can improve model performances are not useful in practice. For example, you may be able to explain sales by using weather data. Yet weather data for the future is not yet known. So while this variable may have an important impact on sales, it cannot be used for improving the forecasting accuracy.

Key Takeaways

- The SARIMAX model allows for adding external variables to the SARIMA model.

- SARIMAX has two types of variables:

  - *Endogenous*: The target variable

  - *Exogenous*: The external variables

- You need to specify the future values of the exogenous variable when forecasting. Therefore, when using exogenous variables, it is important to know that you have fixed information for the future about them.

  - A variable like holidays can work, as you know which holidays will occur in the future.

  - A variable like weather cannot work, as you will not know which weather will occur in the future.

## CHAPTER 9

The VAR Model

In this chapter, you will discover the VAR model, short for the **Vector Autoregression model**. Vector Autoregression models the development over time of multiple variables at the same time.

The term autoregression should sound familiar from the previous chapters. The Vector Autoregression model is part of the family of time series models. Like other models from this category, it predicts the future based on developments in the past of the target variables.

There is an important specificity to Vector Autoregression. Unlike most other models, Vector Autoregression models multiple target variables at the same time. The multiple variables are used at the same time as target variables and as explanatory variables for each other. A model that uses multiple target variables in one model is called a **multivariate model** or, more specifically, **multivariate time series**.

The Model Definition

Since the VAR model proposes one model for multiple target variables, it regroups those variables as a vector. This explains the name of the model. The model definition is as follows:

$$y_t = c + A_1 y_{t-1} + A_2 y_{t-2} + \cdots + A_p y_{t-p} + e_t$$

In this formula

- y_t is a vector that contains the present values of each of the variables.

- The order of the VAR model, **p**, determines the number of time steps back that are used for predicting the future.

- c is a vector of constants: one for each target variable.

© Joos Korstanje 2021
J. Korstanje, *Advanced Forecasting with Python*, https://doi.org/10.1007/978-1-4842-7150-6_9

- There is a vector of coefficients *A* for each lag.

- The error is denoted as *e*.

Order: Only One Hyperparameter

So how does the number of lags work in the VAR model? In the previous models, you have seen relatively complex situations with hyperparameters to be estimated for the order of different types of processes.

The VAR model is much simpler in this regard. There is only one order to be defined: the order of the model as a whole. The notation for this order is **VAR(p)**, in which p is the order.

The order defines how many time steps back in time are taken into account for explaining the present. An order 1 means only the previous time step is taken into account. An order 2 means that the two previous time steps are used. And so on. You can also call this the number of lags that are included.

If a lag is included, it is always included for all the variables. It is not possible with the VAR model to use different lags for different variables.

Stationarity

Another concept that you have seen before is stationarity, so let's find out how that works in the VAR model.

If you remember, stationarity means that a time series has no trend: it is stable over the long term. You can use the Augmented Dickey Fuller test to test whether a time series is stationary or not.

You have also seen the technique called differencing, or integration, to make a non-stationary time series stationary.

This theory is also very important for the VAR model. A VAR model can work only if *each of the variables in the model is stationary*. When doing VAR models, you generally work on multiple variables at the same time, so it can be a bit cumbersome to handle if some of the time series are stationary and others not. In that case, differencing should be applied only to the non-stationary time series.

There is one exception to this rule: the Vector Error Correction model is a type of VAR that does not require stationarity. This model can be a good alternative to the VAR model, so I advise you to have a look at it over here: `www.statsmodels.org/stable/generated/statsmodels.tsa.vector_ar.vecm.VECM.html`.

Estimation of the VAR Coefficients

The coefficients of the VAR model can be estimated using a technique called Multivariate Least Squares. The fact that the VAR model uses this technique makes it computationally relatively efficient. To understand this estimation technique, let's pose the VAR model in matrix notation:

$$Y = BZ + U$$

In this case

- Y is a matrix that contains the y values for each variable (each row) and each lag (each column) for future lags.
- B is the coefficient matrix.
- Z is a matrix with the past lags of the y variables.
- U is a matrix with the model errors.

The Multivariate Least Squares allows you to compute the coefficients using the following matrix computation:

$$\hat{B} = YZ'\left(ZZ'\right)^{-1}$$

The coefficients that are estimated this way are what's underlying the model that you'll see later on.

One Multivariate Model vs. Multiple Univariate Models

Vector Autoregression should be applied to multiple target variables that are correlated. If there is no or very little correlation between the variables, they cannot benefit from being combined in one and the same model.

Besides using a VAR model only in cases where it makes sense to combine variables in one and the same model, it is even more important to use objective model evaluation techniques using a train-test set and cross-validation. This can help you make the right choice between using one model for multiple variables and using a separate univariate model for each variable.

An Example: VAR for Forecasting Walmart Sales

In this example, you will continue to work on the Walmart sales data that have been presented in the previous chapter. Yet in the current example, rather than summing the sales per week, you will sum the weekly data per store. As there are 45 stores in the dataset, this yields 45 weekly time series. You can create this data with a plot using Listing 9-1.

Listing 9-1. Preparing the Walmart data per store

```
import pandas as pd
import matplotlib.pyplot as plt
data = pd.read_csv('walmart/train.csv')
data = data.pivot_table(index = 'Date', columns = 'Store',
values = 'Weekly_Sales')

ax = data.plot(figsize=(20,15))
ax.legend([])
ax.set_ylabel('Sales')
plt.show()
```

This code will give you the plot that is shown in Figure 9-1. As you can see in this plot, the data per store follows the same pattern. For example, you can see that they almost all peak at the same moment. This shows the interest in using multivariate time series.

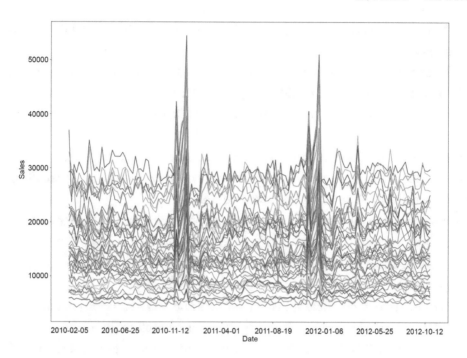

Figure 9-1. *The plot of the 45 stores*

Now let's get into the fit of a VAR model. As before, let's use a ten-step forecast on a ten-step test dataset. The code to create a VAR model can be seen in Listing 9-2.

Listing 9-2. Fitting the VAR model

```
from sklearn.metrics import mean_absolute_percentage_error
from statsmodels.tsa.api import VAR

train = data.iloc[:-10,:]
test = data.iloc[-10:,:]

model = VAR(train)
results = model.fit(maxlags=2)

lag_order = results.k_ar
fcst = results.forecast(train.values[-lag_order:], 10)

model_accuracy = 1 - mean_absolute_percentage_error(test, fcst)
print(model_accuracy)
```

There are a few interesting things to notice in this code, which are different from what you have seen before.

Firstly, note that the data on which the model is fit is a multivariate dataset: it uses all the columns at once. This is as could be expected from a multivariate time series model.

Secondly, note that rather than doing a grid search, the example shows the use of an argument called **maxlags**. This means that the VAR model will optimize the choice for the order of the model itself, all while respecting a maximum lag. In this case, the maxlags has been chosen at two, as going above two would require more coefficients to be estimated than possible using the current data.

The maxlags chooses the ideal order of the model using the **AIC**, the **Akaike Information Criterion**. The Akaike Information Criterion is a famous KPI for goodness of fit of a model. The AIC is based only on the training data. It is therefore more prone to inducing overfitted models.

AIC is more used in classical statistical models and less in modern machine learning. Yet it is a fast and practical way of choosing order and therefore important to now. The negative aspect of VAR is that it requires very many parameters to be estimated. This makes it **require enormous amounts of data** to fit models with higher orders. In the current example, it is impossible to use an order higher than two.

Also, you should notice in the code that when applying the forecast method, it is necessary to give the last part of training data. This is needed so that the model can compute the future by applying coefficients to the lagged variables.

The model accuracy obtained with the model is **0.89**.

Despite the low possibility of model tuning with this model, it must be said that the performances are relatively good. But as always, remember that in practice, the only way to know whether this is good enough is by doing benchmarking and model comparison with other models.

Key Takeaways

- The VAR model uses multivariate correlation to make one model for multiple target variables.

- The order of the VAR model, p, determines the number of time steps back that are used for predicting the future.

- The VAR model implementation can define the ideal number of lags using the maxlags parameter and the Akaike Information Criterion.

- The VAR model needs to estimate a large number of parameters, which makes it require a huge amount of historical data. This makes it difficult to estimate higher lags.

CHAPTER 10

The VARMAX Model

You have discovered the VAR model in the previous chapter. And just like in univariate time series, there are some building blocks that can be built upon the VAR model to account for different types of processes. This can build up from the smaller and more common **VAR** to the more complex **VARMAX**.

It must be said that techniques for multivariate time series modeling are a part of the more advanced techniques. They do not always have Python implementations and are less often used than the techniques for univariate time series. While the use of the VAR model is still relatively common, the more advanced models in this branch become less and less documented on the Internet.

In this chapter, you will discover the VARMAX model. It is the go-to model for multivariate time series. It adds a **moving average component** to the VAR model, and it can allow for external, or exogenous, variables as well. The components in the VARMAX model are therefore

- **V** for vector indicating that it's a multivariate model

- **AR** for autoregression

- **MA** for moving average

- **X** for the use of exogenous variables (in addition to the endogenous variables)

It must be noted that the VARMAX model does *not* have a seasonal component. This is not necessarily a problem, as seasonality can be included through the *exogenous variables*. For example, a monthly seasonality can be modeled by adding the exogenous variable month. For a weekly seasonality, you could add the variable week number.

Another missing block is the integration block. The VARMAX model does not include the differencing of non-stationary time series. As for the VAR model, all the input time series must already be *stationary*.

© Joos Korstanje 2021
J. Korstanje, *Advanced Forecasting with Python*, https://doi.org/10.1007/978-1-4842-7150-6_10

Model Definition

The mathematical model of the VARMAX definition is as follows:

$$y_t = v + A_1 y_{t-1} + \cdots + A_p y_{t-p} + Bx_t + \in_t + M_1 \in_{t-1} + \cdots + M_q \in_{t-q}$$

In this model, y_t is a vector of the values of the present, and the other y's are the lagged values. The **As** are the autocorrelation coefficients: for each lag, they are a vector of the same length as the number of time series. **Ms** are vectors of coefficients for the lagged model errors. They represent the moving average part of the model.

As the VARMAX model is merely a combination of building blocks that you have seen throughout the previous seven chapters, I will spare you the repetition over here. Don't hesitate to go back to the previous chapters for more details on the intuition and explanations behind different building blocks. Let's now dive into the Python implementation and the example.

Multiple Time Series with Exogenous Variables

As a first step, let's prepare the Walmart data on a by-store and by-week basis. As you have seen in the previous chapter, this dataset lends itself perfectly to multivariate time series modeling.

Listing 10-1. Prepare the Walmart data for the VARMAX model

```
import pandas as pd
import matplotlib.pyplot as plt
data = pd.read_csv('walmart/train.csv')
exog = data.groupby('Date')['IsHoliday'].sum() > 0
exog = exog.apply(lambda x: float(x))

data = data.pivot_table(index = 'Date', columns = 'Store',
values = 'Weekly_Sales')

ax = data.plot(figsize=(20,15))
ax.legend([])
ax.set_ylabel('Sales')
plt.show()
```

Note that in Listing 10-1, the exogenous data have been created as well. They are the indicator of the presence of holidays per week. They will be the only variable used as exogenous data in this example, but there is no theoretical restriction to use more exogenous variables.

The plot obtained by this code is shown in Figure 10-1. You can observe that the time series follow patterns that seem relatively correlated.

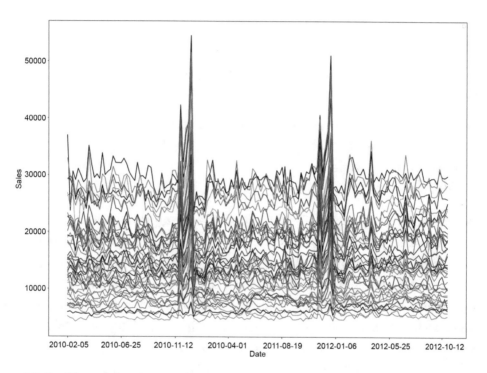

Figure 10-1. *Plot of the time series per store*

In the previous chapter, you have estimated the VAR, using only the order p. This p determines the order of the AR component. In the VARMAX model, there are both an AR component and an MA component. Therefore, there is an order for the AR part and for the MA part to be decided. The hyperparameters of VARMAX(p,q) are

- The order of the AR component, denoted p

- The order of the MA component, denoted q

Now to estimate a VARMAX model, you may use the code shown in Listing 10-2. To avoid the model taking too much computation resources on your hardware, the code example uses the first three stores of the dataset.

Listing 10-2. Running the VARMAX(1,1) model

```
import statsmodels.api as sm
from sklearn.metrics import mean_absolute_percentage_error

train = data.iloc[:-10,[0,1,2]]
test = data.iloc[-10:,[0,1,2]]

train_exog = exog[:-10]
test_exog = exog[-10:]

mod = sm.tsa.VARMAX(train, order=(1,1), exog=train_exog)
res = mod.fit(maxiter=100, disp=False)

fcst = res.forecast(exog=test_exog.values, steps=10)
mape = mean_absolute_percentage_error(test, fcst)
model_accuracy = 1 - mape
print(model_accuracy)
```

The VARMAX model, especially when applied to examples with a large number of variables, can be very long to run. What's taking long here is the estimation of the moving average part of the model. MA models are known to be exceptionally slow to fit, and in the current case, there is the difficulty of having 45 variables to estimate it for.

The R2 that was obtained on this example was **0.96**. This is a great score. This was obtained on a **ten-step forecast**. As you may remember from Chapter 4, the moving average component is *not suited for multistep forecasts*.

With a model training time that is so long already, it would hardly be a solution to add repetitive retraining procedures to the model. The VARMAX model should be used only in cases where training times are not a problem or where the VARMAX obtains performances that cannot be obtained using alternative training methods.

It is great to have the more advanced modeling technique of VARMAX in your forecasting toolbox. The model can fit more complex processes than many other time series models. Yet it also has its disadvantages: training times are relatively long compared to simpler models, and it needs a relatively large amount of data to estimate correctly.

As a sidenote, you will find that those training time and data availability requirements are almost unavoidable for any of the more complex time series and machine learning techniques. This can partly be countered by computing power, but it is still important to evaluate model choice critically.

Key Takeaways

- The VARMAX model consists of

    - V for vector: it is a multivariate model as it models multiple time series at the same time.

    - AR for autoregression.

    - MA for moving average.

    - X for the addition of exogenous variables.

- The VARMAX(p,q) model takes two hyperparameters:

    - p for the order of the AR part

    - q for the order of the MA part

- The time series in a VARMAX have to be stationary.

PART IV

Supervised Machine Learning Models

CHAPTER 11

The Linear Regression

In the following chapters, you will see the most common supervised machine learning models. As you'll remember from Chapter 1, supervised machine learning algorithms work differently than time series models.

In supervised machine learning models, you try to identify relations between different variables:

- *Target variable*: The variable that you try to forecast

- *Explanatory variables*: Variables that help you to predict the target variable

For forecasting, it is important to understand which types of explanatory variables you can or cannot use. As an example, let's say that the sales of hot chocolate strongly depend on the temperature. When the weather is cold, sales are high. When the weather is warm, sales are low.

You could make a model that regroups this basic if/else logic. Yet, when you think about it, this logic could not be used for forecasting in the future. After all, if you want to forecast tomorrow's sales of hot chocolate, you must know tomorrow's temperature. And this is not something you know: it would require an additional forecast of temperature!

Another example is to predict hot chocolate sales not based on the weather, but on the week number. You could try to identify a relationship with the week number based on past data. Then to predict future sales, you would input the next week's week number into the model and obtain a forecast. This is possible here because the week number is something that you know for certain in advance.

This is important to keep in mind when doing supervised models in general. Now, let's get into more depth in linear regression.

© Joos Korstanje 2021
J. Korstanje, *Advanced Forecasting with Python*, https://doi.org/10.1007/978-1-4842-7150-6_11

The Idea Behind Linear Regression

The idea behind linear regression is to define a linear relationship between a target variable and numerous explanatory variables to predict the target variable. Linear regression is widely used, not only for forecasting. Like any supervised model, as long as you put explanatory variables of the future as input to the model, this works perfectly.

Model Definition

Linear regression is defined as follows:

$$y = \beta_0 + \beta_1 x_1 + \cdots + \beta_p x_p + \varepsilon$$

In this formula

- There are **p** explanatory variables, called **x**.

- There is one target variable called **y**.

- The value for y is computed as a constant (β_0) plus the values of the x variables multiplied by their coefficients β_1 to β_p.

Figure 11-1 shows how to interpret B0 and B1 visually. It shows that for an increase of 1 in the x variable, the increase in the y variable represents β_1. β_0 is the value for y when x is 0.

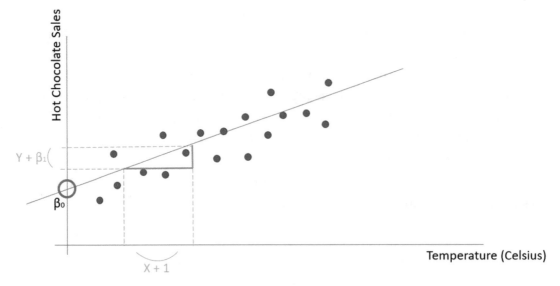

Figure 11-1. *Visual interpretation of linear regression*

To be able to use linear regression, you need to estimate the coefficients (betas) on a training dataset. The coefficients can then be estimated using the following formula, in matrix notation:

$$\hat{\beta} = \left(X^T X\right)^{-1} X^T y$$

This formula is known as **OLS**: the **Ordinary Least Squares method**. This model is very fast to fit, as it requires only matrix calculations to compute the betas. Although easy to fit, it is less suited for more complex processes. After all, it is a linear model, and it can therefore only fit linear processes.

A linear model can fit any type of relationship that goes in one direction. An example of this is "if the temperature goes up, hot chocolate sales go down." Yet linear models cannot fit anything nonlinear. An example of a nonlinear process is this: "if the temperature is below zero, hot chocolate sales are low; if the temperature is between zero and 10, hot chocolate sales are high; if the temperature is high, hot chocolate sales are low."

As you see, the second example is nonlinear because you could not draw a straight line from low to high hot chocolate sales. Rather, you could better make an if/else statement to capture this logic.

You should keep in mind that linear models are not very good at capturing nonlinear trends. Yet nonlinear models, when tuned correctly, can often approximate linear trends quite well. This is the reason that many of the more advanced machine learning techniques use a lot of nonlinear approaches. You will see this throughout the following chapters.

Example: Linear Model to Forecast CO$_2$ Levels

As an example for the linear model, you will work with the CO$_2$ dataset that you have already discovered in a previous chapter. This dataset has the advantage of being relatively easy to work with, and it will be a perfect case to show how to add **lagged variables, seasonality, and trend** into a supervised machine learning model.

Let's get started by importing the data into Python and plotting it. The code for this is shown in Listing 11-1.

Listing 11-1. Importing the data and plotting it

```
import statsmodels.api as sm
import pandas as pd
import matplotlib.pyplot as plt

data = sm.datasets.co2.load_pandas()
co2 = data.data
co2 = co2.dropna()
ax = co2.plot()
ax.set_ylabel('CO2 level')
plt.show()
```

Figure 11-2 shows the graph that you should obtain.

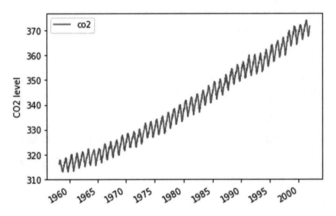

Figure 11-2. CO_2 levels over time

Now, you only have the dates and the CO_2 values. The interesting step here is to do **feature engineering**: creating additional variables, based on the original variables. Even though there is very little information in this dataset, there are a lot of variables that you can create from it.

Let's start by extracting seasonal variables from the date variable. As we can see from the plot, there is a strong seasonal pattern going on. You could try to capture this by adding a monthly seasonality to the model. For this, it is necessary to create a variable month in your dataset. You can use Listing 11-2 to do this.

Listing 11-2. Creating the variable month

```
co2['month'] = [x.month for x in co2.index]
```

Now that you have this variable for monthly seasonality, let's see whether you can create a variable that captures the long-term upward trend. The solution to this problem is to add a variable year, by extracting the year from the date variable. As there is a yearly increase in the data, the trend effect could be captured by this variable. This is done in Listing 11-3.

Listing 11-3. Creating the variable year

```
co2['year'] = [x.year for x in co2.index]
```

Now for starters, let's just try to fit a linear regression with only those two explanatory variables month and year. The package scikit-learn, which you have seen before, contains a large number of supervised models and will be used for this exercise. This basic model is created in Listing 11-4.

Listing 11-4. Fitting a linear regression with two variables

```
# Create X and y objects
X = co2[['year', 'month']]
y = co2['co2']

# Create Train test split
from sklearn.model_selection import train_test_split
X_train, X_test, y_train, y_test = train_test_split(X, y, test_size=0.20,
random_state=12345,shuffle=False)

# Fit model
from sklearn.linear_model import LinearRegression
from sklearn.metrics import r2_score

my_lm = LinearRegression()
my_lm.fit(X = X_train, y = y_train)

train_fcst = my_lm.predict(X_train)
test_fcst = my_lm.predict(X_test)
```

```
train_r2 = r2_score(y_train, train_fcst)
test_r2 = r2_score(y_test, test_fcst)

print(train_r2, test_r2)

# Plot result
plt.plot(list(test_fcst))
plt.plot(list(y_test))
plt.xlabel('Steps into the test set')
plt.ylabel('CO2 levels')
plt.show()
```

The train R2 of this model is **0.96**, which is great. The test R2, however, is **0.34**, which is relatively bad. In Figure 11-3, you can see the fit on the test set and see that there is some improvement to be made.

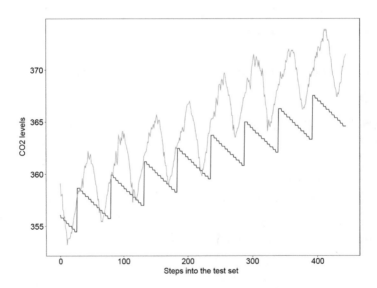

Figure 11-3. *Predictive performance plot of the simple model*

The next thing that you will add to the model is an autoregressive component. This will be done by feature engineering. As always in supervised models, you need to extract any information as explanatory variables. For this example, let's see how to use the shift method to create a lagged variable. You can, for example, add five lagged variables easily using Listing 11-5.

Listing 11-5. Adding lagged variables into the data

```
co2['co2_l1'] = co2['co2'].shift(1)
co2['co2_l2'] = co2['co2'].shift(2)
co2['co2_l3'] = co2['co2'].shift(3)
co2['co2_l4'] = co2['co2'].shift(4)
co2['co2_l5'] = co2['co2'].shift(5)
```

Note that adding lagged variables creates NA in the dataset. This is a border effect, and it is not a problem. Simply delete any missing data with Listing 11-6.

Listing 11-6. Drop missing values

```
co2 = co2.dropna()
```

As a final step, let's fit and evaluate the model with the monthly seasonality, the yearly trend, and the five autoregressive lagged variables. You must note here that, since we added the lagged values into the train set, it would be impossible to do this for multiple steps forward. You calculate the first future value using the data of today. You calculate the second future value by using the data of tomorrow. Therefore, the error that you evaluate here should be interpreted as a one-step forecasting error, whereas the previous code block did a multistep forecast. This is done in Listing 11-7.

Listing 11-7. Fitting the full linear regression model

```
# Create X and y objects
X = co2[['year', 'month', 'co2_l1', 'co2_l2', 'co2_l3', 'co2_l4',
'co2_l5']]
y = co2['co2']

# Train Test Split
X_train, X_test, y_train, y_test = train_test_split(X, y, test_size=0.20,
random_state=12345,shuffle=False)

# Fit the model
my_lm = LinearRegression()
my_lm.fit(X = X_train, y = y_train)

train_fcst = my_lm.predict(X_train)
test_fcst = my_lm.predict(X_test)
```

```
train_r2 = r2_score(y_train, train_fcst)
test_r2 = r2_score(y_test, test_fcst)

print(train_r2, test_r2)

# Plot result
plt.plot(list(test_fcst))
plt.plot(list(y_test))
plt.xlabel('Steps into the test set')
plt.ylabel('CO2 levels')
plt.show()
```

The R2 train score of this model is **0.998**, and the test R2 score is **0.990**. This is a great performance! You can see in the plot (Figure 11-4) that the predictions follow the actual values almost perfectly.

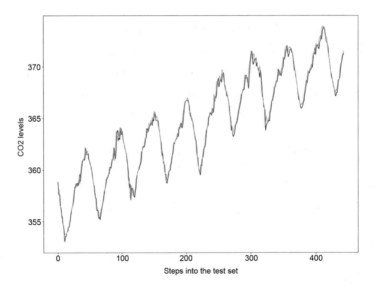

Figure 11-4. *Predictive performance plot of the full model*

In conclusion, you have seen how to add time series trends into a supervised model. You have used a linear model to obtain this result. In the following chapters, you'll discover how to apply more and more complex models in the same way.

Key Takeaways

- The linear model is the simplest supervised machine learning model.

- The linear model finds the best linear combination of external variables to forecast the future.

- Linear models cannot easily adapt to nonlinear situations.

- Feature engineering is the task of creating the best possible variables in your dataset, in order to obtain explanatory variables that can help your model to obtain best performances.

- Seasonality can be fitted by supervised models when you introduce a seasonal variable.

- A trend can be fitted by supervised models when you introduce a trend variable.

- Autoregression can be fitted by supervised models when you add lagged versions of the target variable into the explanatory variables.

CHAPTER 12

The Decision Tree Model

As you've discovered in the previous chapter, there is a distinction in supervised machine learning models between linear and nonlinear models. In this chapter, you will discover the **Decision Tree** model. It is one of the simplest nonlinear machine learning models.

The idea behind the Decision Tree model can be intuitively understood as a long list of *if-else statements*. Those if-else decisions would be used at the prediction stage: the model predicts some **result x** if a certain condition is true, and it will **predict y** otherwise. As you see, there is no linear trend in this type of logic, and because of this, *a decision tree model can fit nonlinear trends.*

Let's see an example of this decision tree logic in practice, without yet discussing where the decision tree comes from. Figure 12-1 shows an example decision tree that can help you to understand intuitively how a decision tree could work. It shows a decision tree that forecasts the average rainfall depending on climate zone and season.

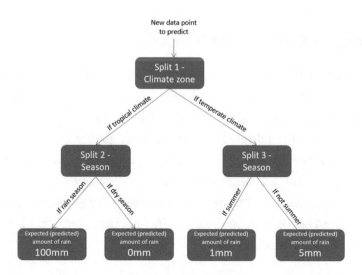

Figure 12-1. *An intuitive example of a decision tree*

© Joos Korstanje 2021
J. Korstanje, *Advanced Forecasting with Python*, https://doi.org/10.1007/978-1-4842-7150-6_12

Mathematics

You'll now discover the mathematical and algorithmic components of the decision tree algorithm. There are different ways to fit decision trees, but the most common one is to have two steps: splitting (also called growing) the tree and pruning the tree.

Splitting

Now that you have seen that the decision tree is merely a hierarchical ordination of multiple decision splits, the logical question is where those decisions come from.

It all starts with a dataset in which you have a target variable and multiple explanatory variables. Now a huge number of splits are possible, but you need to choose one. The first split should be the split that would obtain **the lowest Mean Squared Error**. In the end, even a model with only one split would be able to be used as a predictive model.

Now that you know the goal of your split, the only thing left to do is to test each possible split and evaluate which of them results in the lowest Mean Squared Error. Once you've identified the best first split, you obtain two groups. You then repeat this procedure for each of the two groups, so that you then obtain four groups and so forth.

At some point, you cannot split any further. This point is whenever there is only one data point in each group. Logically, a single data point cannot be split.

Pruning and Reducing Complexity

Although it is possible to let the tree continue splitting until further splitting is impossible, it is not necessarily the best thing to do. There is also a possibility to allow having a bit more data points in a group to avoid overfitting.

This can be obtained in multiple ways. Some implementations allow a pruning process, which first forces a tree to split everything completely and then adds the pruning phase to cut off those branches that are the least needed.

Other implementations let you simply add a complexity parameter that will directly prevent the trees from becoming too detailed. This is a parameter that is a great choice for optimizing using a grid search.

Example

In this chapter, you'll be working with data that come from a bike-sharing company. You can download the data from the UCI machine learning archives over here: `https://archive.ics.uci.edu/ml/datasets/bike+sharing+dataset`.

This will be a difficult example. In this dataset, you'll be predicting the number of rental bike users per day. And as you can imagine, this will depend strongly on the weather. As I explained in the previous chapter, the weather is hard to predict, and if your forecast has a strong correlation with the weather, you know it will be a big challenge.

Let's first have a look at the data to know what you're working with here. This is done in Listing 12-1 and will obtain the plot in Figure 12-2.

Listing 12-1. Import the bike data

```
import pandas as pd
import matplotlib.pyplot as plt
data = pd.read_csv('bikedata/day.csv')
ax = data['cnt'].plot()
ax.set_ylabel('Number of users')
ax.set_xlabel('Time')
plt.show()
```

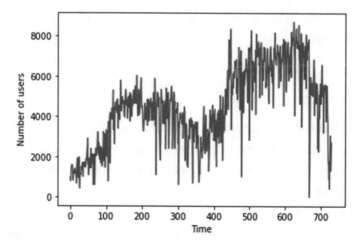

Figure 12-2. *Plot of the bike-sharing users*

As you can see in the plot, many things are going on here. There is a difficult trend to be estimated. There is also a lot of day-to-day variation. When you look at the dataset, you can see that there are a lot of data on weather. This is going to be the challenge: the number of bike-sharing users depends strongly on the day's weather, but the weather is not something that you can know in advance.

The only thing you can do in such cases is to try and do a lot of feature engineering to create many valuable variables for the model. In the code block in Listing 12-2, you will use some of the existing variables, and you'll see how to create a number of variables. The total list of explanatory values is as follows:

- *Original variable* 'season': (1:spring, 2:summer, 3:fall, 4:winter)

- *Original variable* 'yr': The year

- *Original variable* 'mnth': The month

- *Original variable* 'holiday': Whether the day is a holiday

- *Original variable* 'weekday': The day of the week

- *Original variable* 'workingday': Whether the day is a holiday/ weekend or a working day

- *The 7 last days of* 'cnt': An autoregressive component for the number of users

- *The 7 last days of* 'weathersit': The weather, from 4 (very bad weather) to 1 (good weather)

- *The 7 last days of* 'temperature'

- *The 7 last days of* 'humidity'

You will also **delete an influential outlier value** in the dataset. This is an extreme observation, and it is difficult to understand why it happened. You can see it in Figure 12-2, as a very low peak occurring at index 477. As it can influence the model negatively, it is best to not include it in the training data.

There is another low peak somewhere between 650 and 700, but as this part of the data will be our test set, it would be unfair to do a treatment to it. **Removing outliers from the test set would be cheating,** as this would make our test score higher than reality. The goal is always to have a model error estimate that is as reliable as possible, to anticipate its behavior when applying the model in practice.

Listing 12-2. Creating the training dataset

```
# 7 last days of user count (autoregression)
data['usersL1'] = data['cnt'].shift(1)
data['usersL2'] = data['cnt'].shift(2)
data['usersL3'] = data['cnt'].shift(3)
data['usersL4'] = data['cnt'].shift(4)
data['usersL5'] = data['cnt'].shift(5)
data['usersL6'] = data['cnt'].shift(6)
data['usersL7'] = data['cnt'].shift(7)

# 7 last days of weathersit
data['weatherL1'] = data['weathersit'].shift(1)
data['weatherL2'] = data['weathersit'].shift(2)
data['weatherL3'] = data['weathersit'].shift(3)
data['weatherL4'] = data['weathersit'].shift(4)
data['weatherL5'] = data['weathersit'].shift(5)
data['weatherL6'] = data['weathersit'].shift(6)
data['weatherL7'] = data['weathersit'].shift(7)

# 7 last days of temperature
data['tempL1'] = data['temp'].shift(1)
data['tempL2'] = data['temp'].shift(2)
data['tempL3'] = data['temp'].shift(3)
data['tempL4'] = data['temp'].shift(4)
data['tempL5'] = data['temp'].shift(5)
data['tempL6'] = data['temp'].shift(6)
data['tempL7'] = data['temp'].shift(7)

# 7 last days of humidity
data['humL1'] = data['hum'].shift(1)
data['humL2'] = data['hum'].shift(2)
data['humL3'] = data['hum'].shift(3)
data['humL4'] = data['hum'].shift(4)
data['humL5'] = data['hum'].shift(5)
data['humL6'] = data['hum'].shift(6)
data['humL7'] = data['hum'].shift(7)
```

```
data = data.dropna()
data = data.drop(477)

X = data[['season', 'yr', 'mnth', 'holiday', 'weekday', 'workingday',
           'weatherL1', 'weatherL2', 'weatherL3', 'weatherL4', 'weatherL5',
           'weatherL6', 'weatherL7',
          'usersL1','usersL2', 'usersL3', 'usersL4', 'usersL5', 'usersL6',
          'usersL7',
          'tempL1', 'tempL2', 'tempL3', 'tempL4', 'tempL5', 'tempL6',
          'tempL7',
          'humL1', 'humL2','humL3', 'humL4', 'humL5', 'humL6', 'humL7']]

y = data['cnt']
```

Now, let's move on to the model building. To get started, let's do a model without any hyperparameter tuning. The example is shown in Listing 12-3.

Listing 12-3. Fitting the model

```
# Create Train test split
from sklearn.model_selection import train_test_split
X_train, X_test, y_train, y_test = train_test_split(X, y, test_size=0.20,
random_state=12345, shuffle=False)

from sklearn.tree import DecisionTreeRegressor
my_dt = DecisionTreeRegressor(random_state=12345)
my_dt.fit(X_train, y_train)

from sklearn.metrics import r2_score
print(r2_score(list(y_test), list(my_dt.predict(X_test))))
```

This first, unoptimized model makes for an R2 of the test set of **0.166**. Not great yet, so see what can be done using a grid search. In this grid search, we'll be tuning a few hyperparameters:

- min_samples_split: The minimum number of samples required to split a node. Higher values make for fewer splits. Having fewer splits makes a tree less specific, so this can help to prevent overfitting.

- max_features: The number of features (variables) to consider when searching the best split. When using fewer features, you force the splits to be different from each other, but at the same time, you can be blocking the tree from finding the split that is actually the best split.

- *Error criterion*: As we're optimizing our R2, we would potentially want to use the R2 as an error criterion for the decision of the best splits. Yet it is not available. It can therefore be interesting to try the MSE and the MAE and see which one allows us to optimize the R2 of the complete tree.

There are many more hyperparameters in the DecisionTreeRegressor. You can refer to the scikit-learn page for the DecisionTreeRegressor to find them. Don't hesitate to try out and play with those different parameters. The grid search is shown in Listing 12-4.

Listing 12-4. Adding a grid search

```
from sklearn.model_selection import GridSearchCV

my_dt = GridSearchCV(DecisionTreeRegressor(random_state=44),
                {'min_samples_split': list(range(20,50, 2)),
                 'max_features': [0.6, 0.7, 0.8, 0.9, 1.],
                 'criterion': ['mse', 'mae']},
                scoring = 'r2', n_jobs = -1)

my_dt.fit(X_train, y_train)
print(r2_score(list(y_test), list(my_dt.predict(X_test))))
```

When running this in the example notebook, this created an R2 score of **0.55**. Although this is still not amazing, it is a large improvement from the previous model. To find out which parameters have been chosen, you can get the best parameters using the code in Listing 12-5.

Listing 12-5. Finding the best parameters

```
print(my_dt.best_estimator_)
```

This will show you the best parameters as follows:

```
DecisionTreeRegressor(criterion='mae', max_features=0.8,
min_samples_split=48,random_state=44)
```

Apparently, this combination of error criterion **MAE rather than MSE**, a **max_ features of 0.8**, and a **min_samples_split of 48** is the model with the best predictive performance. Now, let's make a forecast with this model and plot how well it fits, using Listing 12-6.

Listing 12-6. Plotting the prediction

```
fcst = my_dt.predict(X_test)

plt.plot(list(fcst))
plt.plot(list(y_test))
plt.ylabel('Sales')
plt.xlabel('Time')
plt.show()
```

The plot that you'll obtain is the plot shown in Figure 12-3.

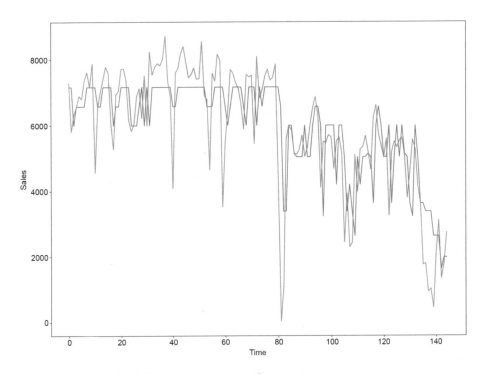

Figure 12-3. *Plot of the bike-sharing users forecast*

A last thing that is very practical with the Decision Tree model is that you can obtain a plot that shows you the different splits that have been identified. This type of plot is sometimes also referred to as a **dendrogram**. Despite the lower performances of a Decision Tree model, one of its strong points is that it allows for interpretation in detail of the decisions of the model. You can do this using Listing 12-7.

Listing 12-7. Plotting the dendrogram

```
from sklearn.tree import plot_tree
plot_tree(my_dt.best_estimator_, max_depth=1)
plt.show()
```

This code will show you the beginning of the decision tree, as shown in Figure 12-4. You can make larger extractions, by increasing the max_depth of the plot.

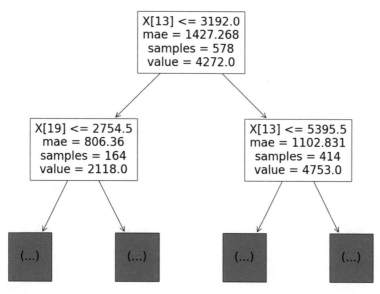

Figure 12-4. *Plot the decision tree*

This type of plot shows one of the real added values of using simpler models like the Decision Tree model. They allow for an interpretation of the model, and this may be required in many circumstances. Models that you'll see in the next chapters will become less and less interpretable.

Key Takeaways

- The Decision Tree model is one of the simplest nonlinear supervised machine learning models.

- You can tune the complexity of the decision tree, which defines how long and complex a decision tree becomes.

- Complex trees risk overfitting: they learn too detailed patterns in the training data. You can use a grid search to optimize the complexity of the tree.

- You can create a dendrogram of the fitted decision tree to obtain all the splits and variables that have been used by your model.

## CHAPTER 13

The kNN Model

In this chapter, you will discover the **kNN model**. The kNN model is the third supervised machine learning model that is covered in this book. Like the two previous models, the kNN model is also one of the simpler models. It is also intuitively easy to understand how the model works. As a downside, sometimes it is not performant enough to compete with the more advanced machine learning models that you will see in the following chapters.

Intuitive Explanation

So what is the kNN model all about? The intuitive idea behind it is simple: you use the data points closest to a new data point to predict it. How could you best predict tomorrow's weather? Look at today's weather! And to predict next month's sales, it is probably reasonably close to this month's sales.

Yet it gets more advanced when considering that a nearest data point can be in multiple dimensions. For example, if next month's sales are in December, the data point could be closest to another month of December in the dimension month.

As soon as you have multiple variables, this idea of distance becomes a very powerful concept for predicting new values. In short, the kNN model tries to find the **nearest neighbors** to a data point and uses their value as a prediction.

Mathematical Definition of Nearest Neighbors

The definition of nearest neighbors is based on the computation of the **Euclidean distance** from the new data point to each of the existing data points. The Euclidean distance is the most common distance measure. You probably use it on a daily basis when talking about your distances from home to work and so on. You can express it mathematically as follows:

© Joos Korstanje 2021
J. Korstanje, *Advanced Forecasting with Python*, https://doi.org/10.1007/978-1-4842-7150-6_13

$$d(p,q) = \sqrt{(p_1 - q_1)^2 + (p_2 - q_2)^2 + \cdots + (p_n - q_n)^2}$$

In this formula, **p** and **q** are two data points each with n dimensions. This would generally mean that there are **n** explanatory variables in the dataset. You sum the squared distances for each dimension and finally take the square root.

The letter **k** is used to indicate the number of neighbors to use. To compute the *k nearest neighbors*, you simply compute the distance between your new data point and each of the data points in the training data. Depending on which number you have for k, you take the k data points that have the lowest distance.

The graph in Figure 13-1 shows how the algorithm works in a two-dimensional situation. All the data points in the training dataset are available to be chosen as a neighbor for any new point. As a dummy example, let's imagine it's again about hot chocolate sales based on the temperature.

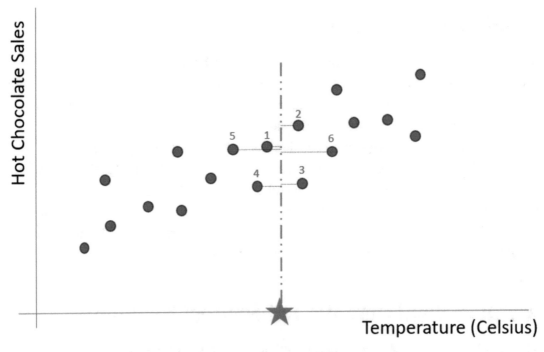

Figure 13-1. *The six nearest neighbors of a new data point*

The blue points in the graph are real observations from the past, in which we observed temperature and hot chocolate prices. Let's say we already know tomorrow's temperature and we want to predict the hot chocolate sales for tomorrow.

If we use the kNN algorithm for this, we need to identify the closest neighbors: in the graph, they are annotated based on distance. The notion of nearest, in this case, is based only on one variable: temperature. In cases with more variables, this would be the same computation using the formula stated earlier, yet it would be hard to visualize such a multivariate situation.

Combining k Neighbors into One Forecast

Once you have identified the **k neighbors** that are closest to your new data point, you do not yet have a prediction. There is one step remaining to convert the multiple neighbors into one prediction. There are two prevalent methods for it:

1. The first method is simply to take **the average** of the target value of the k nearest neighbors. This average is then used as the prediction.

2. The second method is to take **the weighted average** of the k nearest neighbors and use their distances as the inverse weight so that closer points are weighted heavier in the prediction.

Deciding on the Number of Neighbors k

A last thing remains to be decided on, and that is how many nearest neighbors you want to include in the prediction. This is decided by the value of k. To apply this to the previous example, let's see two different cases – one nearest neighbor and three nearest neighbors – and see what the difference in prediction is. The two are given in Figure 13-2.

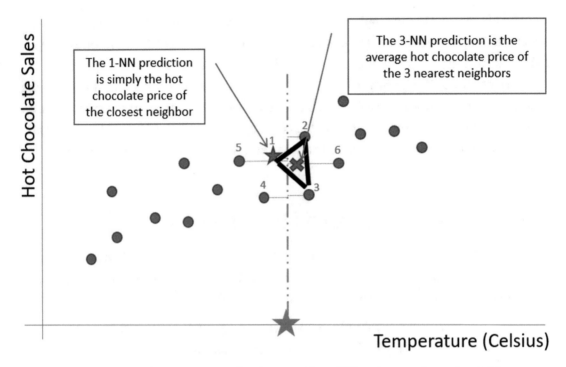

Figure 13-2. *Two alternative predictions with a different number of neighbors used*

The number of neighbors to use, or *k*, is a hyperparameter in the kNN model, and it is best optimized using hyperparameter tuning. The method that you've seen in this book is a grid search CV, which is one of the go-to methods for hyperparameter tuning. In this chapter, you'll also discover an alternative method: random search. Let's first introduce an example to make it more applied.

Predicting Traffic Using kNN

For this example, you'll be working with a dataset of hourly traffic volumes from the Interstate 94. You can find this dataset on the UCI machine learning repository: https://archive.ics.uci.edu/ml/datasets/Metro+Interstate+Traffic+Volume.

If you download this data, you'll find that it has hourly traffic volume, together with some weather data and information about holidays. For the present example, let's avoid depending on weather data and do a forecast based on seasonality and holidays. After all, we could imagine that traffic depends heavily on the time of day, weekdays, and holidays.

To import the data into Python, you can use the code in Listing 13-1.

Listing 13-1. Import the traffic data

```
import pandas as pd
data = pd.read_csv('Metro_Interstate_Traffic_Volume.csv.gz',
compression='gzip')
```

After you've imported this data, let's create the seasonality variables that seem necessary for the modeling exercise for this example. Use the code in Listing 13-2 to create the following variables:

- Year

- Month

- Weekday

- Hour

- IsHoliday

Listing 13-2. Feature engineering to create the additional explanatory variables

```
data['year'] = data['date_time'].apply(lambda x: x[:4])
data['month'] = data['date_time'].apply(lambda x: x[5:7])
data['weekday'] = pd.to_datetime(data['date_time']).apply(lambda x:
x.weekday())
data['hour'] = pd.to_datetime(data['date_time']).apply(lambda x: x.hour)
data['isholiday'] = (data['holiday'] == 'None').apply(float)
```

The next step is to split the data into train and test and fit a default kNN model. This will get you a first feel of the R2 that could be obtained with this type of model. This code is shown in Listing 13-3.

Listing 13-3. Creating the train-test split and computing the R2 of the default model

```
# Create objects X and y
X = data[['year', 'month', 'weekday', 'hour', 'isholiday']]
y = data['traffic_volume']
```

```
# Create Train test split
from sklearn.model_selection import train_test_split
X_train, X_test, y_train, y_test = train_test_split(X, y, test_size=100,
random_state=12345, shuffle=False)

from sklearn.neighbors import KNeighborsRegressor
my_dt = KNeighborsRegressor()
my_dt.fit(X_train, y_train)

fcst = my_dt.predict(X_test)

from sklearn.metrics import r2_score
print(r2_score(list(y_test), list(fcst)))
```

As a positive surprise, the R2 score on this model is very good already: **0.969.** This should mean that the forecast is not far from perfect, so let's try to verify this visually using the code in Listing 13-4.

Listing 13-4. Creating a plot on the data of the test set

```
import matplotlib.pyplot as plt
plt.figure(figsize=(20,20))
plt.plot(list(y_test))
plt.plot(list(fcst))
plt.legend(['actuals', 'forecast'])
plt.ylabel('Traffic Volume')
plt.xlabel('Steps in test data')
plt.show()
```

This code will generate the graph that is shown in Figure 13-3.

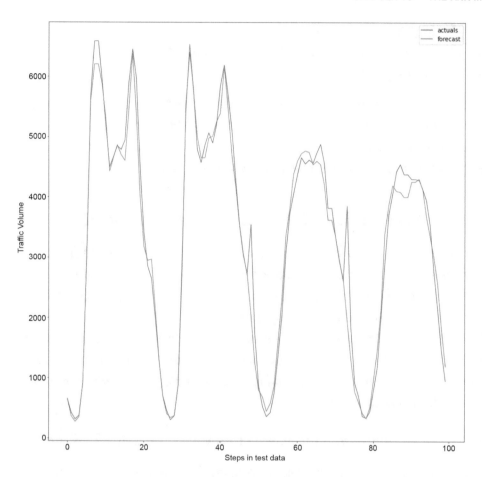

Figure 13-3. *The plot shows the predictive performance on the test dataset*

Grid Search on kNN

The number of neighbors that the scikit-learn implementation uses by default is five. Now as a last step, let's try to add a grid search to see whether a different number of neighbors might get us a better performance. Let's test the number of neighbors by increments of 2, in order to speed it up a little. This is done in Listing 13-5.

Listing 13-5. Adding a grid search cross-validation to the kNN model

```
from sklearn.model_selection import GridSearchCV

my_knn = GridSearchCV(KNeighborsRegressor(),
                {'n_neighbors':[2, 4, 6, 8, 10, 12]},
                scoring = 'r2', n_jobs = -1)
```

```
my_knn.fit(X_train, y_train)
print(r2_score(list(y_test), list(my_knn.predict(X_test))))
print(my_knn.best_estimator_)
```

The R2 score obtained with this tuned model is **0.9695**. This model uses eight nearest neighbors rather than five, and it is better by 0.0002 points on the R2 score. This is very little, and therefore the improvement can hardly be considered existent. In this case, the default model was actually quite good from the start.

Random Search: An Alternative to Grid Search

As a last topic for this chapter, I want to introduce an alternative to grid search. This alternative is called random search. Random search is very easy to understand intuitively: rather than checking every combination of hyperparameters on the grid, it will check a random selection of hyperparameters.

The main reason for this is that it is much faster. Also, it has been found that random search is generally able to obtain results that are very close to those of grid search, and the speed improvement is therefore totally worth it. Listing 13-6 shows how to replace a grid search by a random search.

Listing 13-6. Adding a random search cross-validation to the kNN model

```
from sklearn.model_selection import RandomizedSearchCV

my_knn = RandomizedSearchCV(KNeighborsRegressor(),
                {'n_neighbors':list(range(1, 20))},
                scoring = 'r2', n_iter=10, n_jobs = -1)

my_knn.fit(X_train, y_train)
print(r2_score(list(y_test), list(my_knn.predict(X_test))))
print(my_knn.best_estimator_)
```

The random search has been allowed to choose any number for k between 1 and 20. The argument **n_iter** decides on the number of values that should be **randomly selected** within this range. Note that the selection is random, so this may give a different result than the grid search. There is no way to guarantee that a certain value will be included in the test.

To make a fair comparison with the grid search, we apply a n_iter of 6. As there are six values tested in the grid search, this makes it equivalent. When executing the code, the returned solution is exactly the same as the one returned by grid search.

Key Takeaways

- The kNN model bases its predictions on the values of its nearest neighbors.

- The neighbors' values are combined into a prediction by computing the average or a weighted average based on their distance.

- The Euclidean distance is generally used for the distance measure.

- The number of neighbors to use in the combination is denoted k and is a hyperparameter to the model.

- The value of k can be optimized using hyperparameter tuning, for example, through grid search cross-validation.

- Random search is an alternative to grid search that is faster and generally gives results that are (almost) as good.

CHAPTER 14

The Random Forest

In this chapter, you will discover the **Random Forest model**. It is an easy-to-use model, and it is known to be very performant. The Random Forest and the XGBoost model, which you will discover in the next chapter, are two of the most used machine learning algorithms in modern applications.

A large number of variants on Random Forests and XGBoost exist on the market, yet if you understand the two basics, it will be relatively easy to adapt to any variant.

Intuitive Idea Behind Random Forests

The Random Forest is *strongly based on the Decision Tree model* but adds more complexity to it. As the name suggests, a Random Forest consists of a large number of Decision Trees, each of them with a slight variation.

A Random Forest is much more performant than a Decision Tree. Generally, a Random Forest can combine hundreds or even thousands of Decision Tree models. They will be fitted on slightly different data, as to not be totally equal. So, in short, it's a large number of Decision Trees making predictions that should be close to each other, yet not exactly the same.

Where one machine learning model can sometimes be wrong, the average prediction of a large number of machine learning models is less likely to be wrong. This idea is the foundation of **ensemble learning**.

In the Random Forest, ensemble learning is applied to a repetition of many Decision Trees. Ensemble learning can be applied to any combination of a large number of machine learning models. The reason to use Decision Trees is that it has been proven a performant and easy-to-configure model.

© Joos Korstanje 2021
J. Korstanje, *Advanced Forecasting with Python*, https://doi.org/10.1007/978-1-4842-7150-6_14

Random Forest Concept 1: Ensemble Learning

So how does ensemble learning work exactly? You can understand that having 1000 times the exact same Decision Tree does not have any added value to just using one time this Decision Tree.

In the ensemble model, each of the individual models has to be slightly different from another. There are two famous methods for creating ensembles: **bagging** and **boosting**.

The Random Forest uses bagging to create an ensemble of Decision Trees. In the next chapter, you'll discover the XGBoost algorithm, which uses the alternative technique called boosting.

Let's discover the idea behind bagging. Bagging is short for Bootstrap Aggregation. The idea exists in two parts:

1.   Bootstrap

2.   Aggregation

Bagging Concept 1: Bootstrap

The **bootstrap** is one of the essential parts of the algorithm that makes sure that each of the individual learners fits a slightly different Decision Tree. Bootstrapping means that for each individual learner, the dataset is created by a **resampling process**.

Resampling means creating a new dataset based on the original dataset. The new dataset will be created by randomly selecting data points from the original dataset. Yet the important thing to realize here is that the resampling is done **with replacement**.

Resampling with replacement puts every sampled data point back into the mother population. This makes it possible for a single data point in the original dataset to be selected multiple times in the bootstrapped dataset. There will also be data points in the original data that are not selected for the bootstrapped dataset. A schematic drawing is shown in Figure 14-1.

Original Data Set:
- 8 data points
- Each point is different

Bootstrapped Data Set:
- Still 8 data points
- Some points present multiple times
- Some points not present

Figure 14-1. *The bootstrap*

Bagging Concept 2: Aggregation

The **aggregation** part describes the fact of using multiple learners. In the Random Forest, the bootstrap is executed many times. The exact number is defined by a hyperparameter called **n_estimators**.

For each n_estimators, a Decision Tree is built using a bootstrapped dataset. This will generate a large number of Decision Trees that are all slightly different due to the differences in the datasets. In the end, you can use each of the Decision Trees to make a prediction. However, you would end up with many slightly different predictions.

The solution to this problem is in the aggregation part. To combine all the individual Decision Tree predictions into one Random Forest prediction, you simply take the average of all the individual predictions. It is relatively straightforward, yet a crucial part of the algorithm.

The idea behind this is that the aggregation of many weak learners will result in errors that average each other out. This makes the Random Forest a very performant machine learning algorithm.

Random Forest Concept 2: Variable Subsets

Besides the bootstrapping approach, the Random Forest has a **second process** in place to make sure that the individual learners are not the same. This process does not apply to the data points that are or are not used, but rather **applies to the variables that are or are not used**.

As explained in Chapter 12, the Decision Tree checks each of the variables to find the best split to add to the tree. Yet when using a subset of each variable, you only let the tree check out the splits in a randomly selected number of variables. The exact quantity is defined by a hyperparameter called **mtry** in most mathematical descriptions and called **max_features** in scikit-learn. Although this can sometimes cause a selection of a suboptimal split, it does help to make the individual Decision Trees different from each other.

This value is generally around **80%**, but it can be higher or lower. Together, **n_estimators** and **max_features** are the main hyperparameters of the Random Forest model.

Predicting Sunspots Using a Random Forest

Let's develop an example of the Random Forest model. For this example, you'll again take the sunspot data that you've used in Chapter 5. The ARMA model in Chapter 5 was able to obtain an R2 of **0.84**. This time let's try to forecast the monthly number of sunspots rather than the yearly sum as was done in the previous example. You can import the data using Listing 14-1.

Listing 14-1. Importing the data

```
import pandas as pd
data = pd.read_csv('Ch05_Sunspots_database.csv')
data = data.iloc[:,[1,2]]
```

Now it is necessary to add some feature engineering. Let's add the variables Year and Month to account for seasonality and the lagged versions of the target variable for the past 12 months. This task can be expected to be a bit harder, as there is a more detailed variation that the model needs to learn. The code in Listing 14-2 shows you how this can be done.

Listing 14-2. Feature engineering

```
# Seasonality variables
data['Date'] = pd.to_datetime(data['Date'])
data['Year'] = data['Date'].apply(lambda x: x.year)
data['Month'] = data['Date'].apply(lambda x: x.month)

# Adding a year of lagged data
data['L1'] = data['Monthly Mean Total Sunspot Number'].shift(1)
data['L2'] = data['Monthly Mean Total Sunspot Number'].shift(2)
data['L3'] = data['Monthly Mean Total Sunspot Number'].shift(3)
data['L4'] = data['Monthly Mean Total Sunspot Number'].shift(4)
data['L5'] = data['Monthly Mean Total Sunspot Number'].shift(5)
data['L6'] = data['Monthly Mean Total Sunspot Number'].shift(6)
data['L7'] = data['Monthly Mean Total Sunspot Number'].shift(7)
data['L8'] = data['Monthly Mean Total Sunspot Number'].shift(8)
data['L9'] = data['Monthly Mean Total Sunspot Number'].shift(9)
data['L10'] = data['Monthly Mean Total Sunspot Number'].shift(10)
data['L11'] = data['Monthly Mean Total Sunspot Number'].shift(11)
data['L12'] = data['Monthly Mean Total Sunspot Number'].shift(12)
```

Now that you have got a dataset, let's do a train-test split and fit the Random Forest with the default hyperparameters. Use the code in Listing 14-3 to fit this default model.

Listing 14-3. Fitting the default Random Forest regressor

```
# Create X and y object
data = data.dropna()
y = data['Monthly Mean Total Sunspot Number']
X = data[['Year', 'Month', 'L1', 'L2', 'L3', 'L4', 'L5', 'L6', 'L7', 'L8',
'L9', 'L10', 'L11', 'L12']]

# Create Train test split
from sklearn.model_selection import train_test_split
X_train, X_test, y_train, y_test = train_test_split(X, y, test_size=0.1,
random_state=12345, shuffle=False)

from sklearn.ensemble import RandomForestRegressor
my_rf = RandomForestRegressor()
```

```
my_rf.fit(X_train, y_train)
fcst = my_rf.predict(X_test)

from sklearn.metrics import r2_score
r2_score(list(y_test), list(fcst))
```

Using this model, you will obtain an R2 score of the prediction of **0.863**.

Grid Search on the Two Main Hyperparameters of the Random Forest

This is already a great result, but as always, let's see if that can be optimized using a grid search hyperparameter tuning. This will be shown in Listing 14-4.

Listing 14-4. Fitting the Random Forest regressor with hyperparameter tuning

```
from sklearn.model_selection import GridSearchCV

my_rf = GridSearchCV(RandomForestRegressor(),
                {'max_features':[0.65, 0.7, 0.75, 0.8, 0.85, 0.9, 0.95],
                'n_estimators': [10, 50, 100, 250, 500, 750, 1000]},
                scoring = 'r2', n_jobs = -1)

my_rf.fit(X_train, y_train)
print(r2_score(list(y_test), list(my_rf.predict(X_test))))
print(my_rf.best_params_)
```

This will obtain an R2 score of **0.870**, slightly better than the default version. The optimal hyperparameters that have been identified are max_features (mtry) of **0.7** and n_estimators (ntrees) of **750**.

It would be useful now to have a look at the performance plot to see how it looks like on the test data. This can be done using Listing 14-5, and this will obtain Figure 14-2.

Listing 14-5. Obtaining the plot of the forecast on the test data

```
import matplotlib.pyplot as plt
plt.plot(list(fcst))
plt.plot(list(y_test))
```

```
plt.legend(['fcst', 'actu'])
plt.ylabel('Sunspots')
plt.xlabel('Steps into the test data')
plt.show()
```

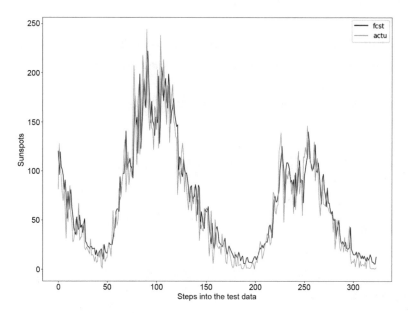

Figure 14-2.   *Predictive plot of sunspots per month*

Random Search CV Using Distributions

As an additional step in this chapter, let's go a bit deeper into the **random search CV** that you discovered in the previous chapter. As you'll understand, the more complicated the models become, the heavier the grid search becomes. Alternatives become more and more interesting. You'll discover more methods for hyperparameter tuning in the next chapters.

In this application of random search, you'll use **distributions** rather than a list of possible values to specify which values you want to test for the different hyperparameters. This is a great functionality of RandomizedSearchCV, especially when you have an idea of what the most likely value for your hyperparameters is.

Let's start by identifying the distributions for max_features and n_estimators and then put them into the RandomizedSearchCV.

Distribution for max_features

For the value of max_features, you previously tested values between 0.65 and 0.95 with steps of 0.05. The most used distribution is the **normal distribution**, so let's see how we could use the normal distribution for this.

You can use the code in Listing 14-6 to try out different normal distributions and check out whether they fit. The normal distribution is defined by a mean and a standard deviation, so those are the two values that you can play around with.

Listing 14-6. Testing out a normal distribution for the max_features

```python
import numpy as np
import scipy.stats as stats
import math

mu = 0.8
variance = 0.005
sigma = math.sqrt(variance)
x = np.linspace(mu - 3*sigma, mu + 3*sigma, 100)
plt.plot(x, stats.norm.pdf(x, mu, sigma))
plt.show()
```

As shown in Figure 14-3, the normal distribution with a mean of 0.8 and a standard deviation of 0.05 covers the range of 0.65–1.0 quite well. The x-axis in this graph shows the value for max_features, and the y-axis shows the corresponding probability for this value being sampled into the random search.

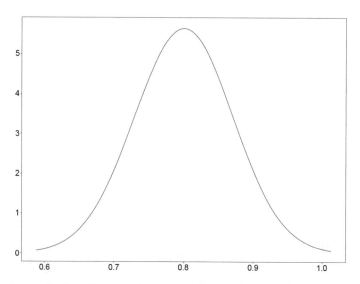

Figure 14-3. *Normal distribution with mean 0.8 and st. dev. 0.05*

As those values seem quite appropriate for max_features, we'll use this distribution in the RandomizedSearchCV.

Distribution for n_estimators

For the distribution of n_estimators, let's try something different. Let's say that we want to test values between 50 and 1000 but that we don't really have a good idea of what the actual value for n_estimators should be.

This means that a **uniform distribution** may be appropriate: in a uniform distribution, you specify a minimum and a maximum, and all values in between are **equally probable** to be selected. This is shown in Listing 14-7.

Listing 14-7. Testing out a uniform distribution for the n_estimators

```
x = np.linspace(0, 2000, 100)
plt.plot(x, stats.uniform.pdf(x, 50, 950))
plt.show()
```

The resulting plot is shown in Figure 14-4.

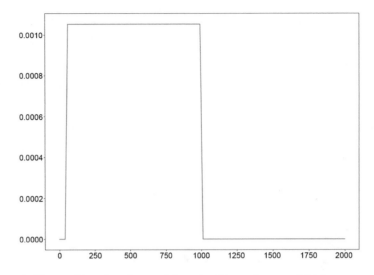

Figure 14-4. *Uniform distribution with min 50 and max 1000*

Yet there is a small problem. The uniform distribution returns a floating number, while the n_estimators can only take an **integer**. The better alternative for a uniform integer is to use the function **scipy.stats.randint**, which returns a random integer between a minimum and a maximum.

Fitting the RandomizedSearchCV

Let's now fit the RandomizedSearchCV using the code in Listing 14-8.

Listing 14-8. RandomizedSearchCV with two distributions

```
from sklearn.model_selection import RandomizedSearchCV

# Specifying the distributions to draw from
distributions = {
    'max_features': stats.norm(0.8, math.sqrt(0.005)),
    'n_estimators': stats.randint(50, 1000)
}

# Creating the search
my_rf = RandomizedSearchCV(RandomForestRegressor(),

                          distributions, n_iter=10,
```

```
                        scoring = 'r2',
                        n_jobs = -1,
                        random_state = 12345)

# Fitting the search
my_rf.fit(X_train, y_train)

# Printing the results
print(r2_score(list(y_test), list(my_rf.predict(X_test))))
print(my_rf.best_params_)
```

The results that you should obtain are an R2 of **0.86**, and the selected hyperparameters are listed in the following. Be aware that due to randomness, you may obtain slightly different results:

- 'max_features': 0.70

- 'n_estimators': 819

The R2 is slightly lower than the one obtained by the grid search. Yet you should observe a very strong increase in execution time. The RandomizedSearchCV is much faster while only losing a very slight amount of performance: a great argument for using RandomizedSearchCV rather than GridSearchCV.

Interpretation of Random Forests: Feature Importance

As you have understood by now, the Random Forest is a relatively complex model. When making a prediction, it combines the predictions of many Decision Trees. This gives the model great performance, but it also makes the model relatively difficult to interpret.

For the Decision Tree model, you have seen how you can plot the tree and follow its decisions based on a new observation. For the Random Forest, you would need to do this many times, making for a very difficult process.

Luckily, an alternative exists for interpreting the Random Forest. This method is called feature importance.

The feature importance of each variable is computed all the way throughout the fitting of the Random Forest. Every time that a variable is used for a split, the error reduction that is brought about by this split is added to the variable's feature importance. At the end of the algorithm, those added-up errors are standardized so that the sum of the variable importance for all variables is 1.0.

The higher the future importance of a variable, the more important the role it has played in the model and therefore the more predictive value it has for your forecast.

You can obtain the feature importances from a Random Forest using Listing 14-9. Note that the Random Forest here is the one fit with RandomizedSearchCV, which is why you call best_estimator first. You also see how to combine the array of feature importances into a more accessible dataframe.

Listing 14-9. Feature importances

```
fi = pd.DataFrame({
        'feature': X_train.columns,
        'importance': my_rf.best_estimator_.feature_importances_})

fi.sort_values('importance', ascending=False)
```

The result looks as shown in Figure 14-5.

	feature	importance
2	L1	0.575328
3	L2	0.205510
4	L3	0.079136
5	L4	0.041819
6	L5	0.013598
7	L6	0.011738
10	L9	0.010307
12	L11	0.010080
11	L10	0.009442
9	L8	0.009358
8	L7	0.009231
13	L12	0.009185
0	Year	0.009022
1	Month	0.006245

Figure 14-5. *Feature importances of the Random Forest*

As you see, the most important feature here is the L1: the 1-lag autoregressive component. The following importances are the other lags. We can understand from this that the model is mainly learning an autoregressive effect with the more recent lags being more important. This is an interesting learning for understanding the model. In some cases, this information can also help you in improving the model even further by improving the selection of your variables, or it can give you hints on how to improve your feature engineering.

Key Takeaways

- The Random Forest adds bagging to the Decision Tree model.

- Bagging is an ensemble method: it fits the same model many times and takes their average as a prediction.

- The Random Forest has two main hyperparameters:

  - The number of trees to use in the forest

  - The number of randomly selected variables to use in each Decision Tree split

- Random search is a faster way to tune hyperparameters and often gives results that are not much worse than grid search. You can give probability distributions for each hyperparameter to tune, and the random search will do a certain number of draws in those distributions.

- Variable importance is a way to interpret the results of the Random Forest model. It can also hint at how to improve the model further.

## CHAPTER 15

Gradient Boosting with XGBoost and LightGBM

In this chapter, you will discover the **gradient boosting model**. In the previous chapter, you discovered the idea behind ensemble methods. As a recap, ensemble methods make powerful predictions by combining predictions of numerous small, less performant models.

Boosting: A Different Way of Ensemble Learning

Gradient boosting combines numerous small Decision Tree models to make predictions. Of course, those small decision trees are different from each other; else, there wouldn't be any advantage of using a larger number of them.

The important concept to understand here is how those decision trees come to be different from each other. This is achieved by a process called **boosting**. Boosting and bagging, which you've seen in the previous chapter, are the principal two methods of ensemble learning.

Boosting is an **iterative process**. It adds more and more weak learners to the ensemble model in an intelligent way. In each step, the individual data points are weighted. Data points that are already predicted well will not be important for the learner to be added. The new weak learners will therefore *focus on learning the things that are not yet understood and therefore improve the ensemble.*

You can see a schematic overview of the boosting process in Figure 15-1. With this approach, you iteratively fit weak models that are focusing on the parts of the data that are not yet understood. While doing this, you keep all the intermediate weak models. The ensemble model is the combination of all those weak learners.

© Joos Korstanje 2021
J. Korstanje, *Advanced Forecasting with Python*, https://doi.org/10.1007/978-1-4842-7150-6_15

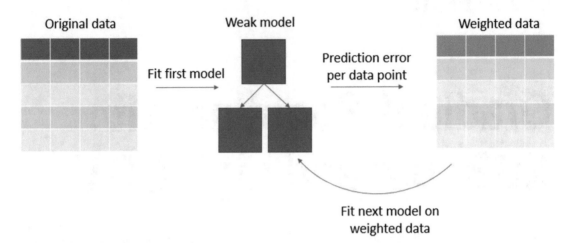

Figure 15-1. *The boosting process*

The Gradient in Gradient Boosting

This iterative process is called *gradient* boosting for a reason. A gradient is a mathematical term that refers to the vector field of partial derivatives that point in the direction of the steepest slope. In simple terms, we often compare gradients to slopes of uphill roads: the higher the slope, the steeper the hill. Gradients are computed by taking derivatives, or partial derivatives, of a function.

In gradient boosting, when adding additional trees to the model, the goal is to add a tree that best explains the variation that was not explained by the previous trees. The target of your new tree is therefore

$$y - \hat{y}$$

This can be denoted rewritten as the negative partial derivative of the loss function against the y predictions:

$$y - \hat{y} = -\frac{\partial L}{\partial \hat{y}}$$

You set this as the target for the new tree to ensure that adding the tree will explain a maximum amount of additional variation in the overall gradient boosting model. This explains why the model is called *gradient* boosting.

Gradient Boosting Algorithms

There are many algorithms that each perform slightly different versions of gradient boosting. When the gradient boosting approach was first invented, the algorithms were not very performant, but that changed with the AdaBoost algorithm: the first algorithm that could adapt to weak learners.

Gradient boosting algorithms are among the most performant machine learning tools on the market. After AdaBoost, a long list of slightly different boosting algorithms has been added to the literature, including XGBoost, LightGBM, LPBoost, BrownBoost, MadaBoost, LogitBoost, and TotalBoost. There are still many contributions being made to improve on gradient boosting theory. In this current chapter, two algorithms will be covered: XGBoost and LightGBM.

XGBoost is one of the most used machine learning algorithms. XGBoost is a quick way to get good performances. As it is easy to use and very performant, it is the first go-to algorithm for many ML practitioners.

LightGBM is another gradient boosting algorithm that is important to know. For the moment, it is a bit less widespread than XGBoost, but it is seriously gaining in popularity. The expected advantage of LightGBM over XGBoost is a gain in speed and memory use.

In this chapter, you will discover the implementations of both of those gradient boosting algorithms.

The Difference Between XGBoost and LightGBM

If you're going to use those two gradient boosting algorithms, it is important to understand in what way they differ. This can also give you an insight into the types of difference that make for such a large number of models on the market.

The difference here is in the way they identify the best splits inside the weak leaners (the individual decision trees). Remember that the splitting in a decision tree is the moment where your tree needs to find the split that most improves the model.

The intuitively simplest idea for finding the best split is to loop through all the potential fits and find the best one. Yet this takes a lot of time, and better alternatives have been proposed by recent algorithms.

The alternative proposed by XGBoost is to use a **histogram-based splitting**. In this case, rather than looping through all possible splits, the model builds histograms of each of the variables and uses those to find the best split per variable. The best overall split is then retained.

LightGBM was invented by Microsoft, and it has an even more efficient method to define the splits. This method is called **Gradient-Based One-Side Sample (GOSS).** GOSS computes gradients for each of the data points and uses this to filter out data points with a low gradient. After all, data points with a low gradient are already understood well, whereas individuals with a high gradient need to be learned better. LightGBM also uses an approach called **Exclusive Feature Bundling (EFB)**, which is a feature that allows for a speed-up when having many correlated variables to choose from.

Another difference is that the LightGBM model fits leaf-wise (best-first) tree growth, whereas XGBoost grows the trees tree-wise. You can see the difference in Figure 15-2. This difference is a feature that would theoretically favor LightGBM in terms of accuracy, yet it comes at a higher risk of overfitting in the case of little data available.

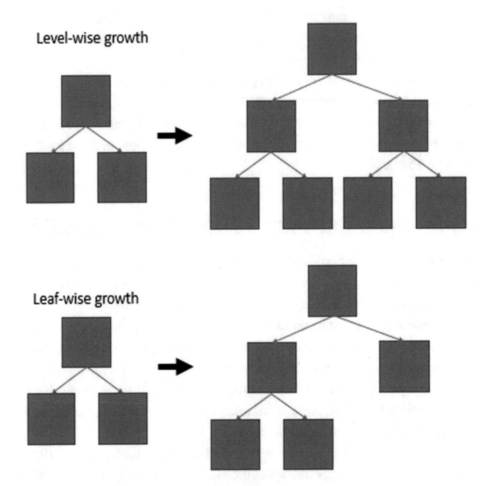

Figure 15-2. *Leaf-wise growth vs. level-wise growth*

For more details on the differences between the gradient boosting algorithms, you can check out the paper by Microsoft (`www.microsoft.com/en-us/research/ publication/lightgbm-a-highly-efficient-gradient-boosting-decision-tree/`).

Forecasting Traffic Volume with XGBoost

For this chapter, we'll be using the same dataset as the one used for the kNN model: interstate traffic. In the previous example, we were able to obtain an R2 of **0.9695**. Gradient boosting is a more performant algorithm, so let's see whether we can use it to improve on this already great score.

You can find this dataset on the UCI machine learning repository: `https://archive. ics.uci.edu/ml/datasets/Metro+Interstate+Traffic+Volume`. You can import the data using Listing 15-1.

Listing 15-1. Importing the data

```
import pandas as pd
data = pd.read_csv('Metro_Interstate_Traffic_Volume.csv')
```

To make a fair benchmark between the gradient boosting models and the kNN model, let's do the exact same feature engineering as the one applied in Chapter 13. This feature engineering is shown in Listing 15-2.

Listing 15-2. Applying the same feature engineering as done previously for the kNN model

```
data['year'] = data['date_time'].apply(lambda x: int(x[:4]))
data['month'] = data['date_time'].apply(lambda x: int(x[5:7]))
data['weekday'] = pd.to_datetime(data['date_time']).apply(lambda x:
x.weekday())
data['hour'] = pd.to_datetime(data['date_time']).apply(lambda x: x.hour)
data['isholiday'] = (data['holiday'] == 'None').apply(float)
```

Let's now do a first test with XGBoost, using Listing 15-3. You use the XGBoost package for this.

Note   Installing libraries using Python can sometimes be a pain due to the need for different dependencies. If you have trouble installing those boosting libraries, you could check out Google Colaboratory notebooks (`https://colab.research.google.com`) or Kaggle notebooks (kaggle.com/notebooks), which are free notebook environments that have all the modern libraries for machine learning at the ready for you.

Listing 15-3. Applying the default XGBoost model

```
# Create objects X and y
X = data[['year', 'month', 'weekday', 'hour', 'isholiday']]
y = data['traffic_volume']

# Create Train test split
from sklearn.model_selection import train_test_split
X_train, X_test, y_train, y_test = train_test_split(X, y, test_size=100,
random_state=12345, shuffle=False)

from xgboost import XGBRegressor
my_xgb = XGBRegressor()
my_xgb.fit(X_train, y_train)

xgb_fcst = my_xgb.predict(X_test)

from sklearn.metrics import r2_score
print(r2_score(list(y_test), list(xgb_fcst)))
```

This results in an R2 score of **0.920.** This is quite less good than the 0.9695 that was obtained using kNN.

Forecasting Traffic Volume with LightGBM

Now let's do the same thing with LightGBM and see how that compares using
Listing 15-4.

Listing 15-4. Applying the default LightGBM model

```
from lightgbm import LGBMRegressor
my_lgbm = LGBMRegressor()
my_lgbm.fit(X_train, y_train)

lgbm_fcst = my_lgbm.predict(X_test)

print(r2_score(list(y_test), list(lgbm_fcst)))
```

The R2 score of LightGBM is **0.972.** This is much better than the default XGBoost
model, and it is also better than the tuned kNN model. Let's create a graph to see where
the XGBoost model and the LightGBM model predicted something different. You can
use Listing 15-5 to create the graph. The graph should be the one in Figure 15-3.

Listing 15-5. Create a graph to compare the XGBoost and LightGBM forecast to
the actuals

```
import matplotlib.pyplot as plt
plt.figure(figsize=(20,10))
plt.plot(list(y_test))
plt.plot(list(xgb_fcst))
plt.plot(list(lgbm_fcst))
plt.legend(['actual', 'xgb forecast', 'lgbm forecast'])
plt.ylabel('Traffic Volume')
plt.xlabel('Steps in test data')
plt.show()
```

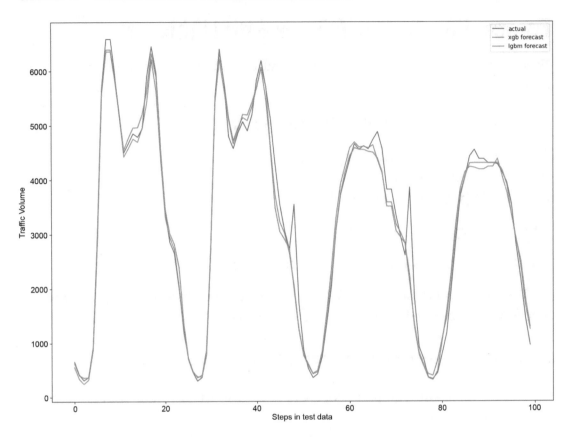

Figure 15-3. *Comparing the two gradient boosting models*

As you can see in this graph, the difference in the two model performances is striking: XGBoost is seriously off at certain places. This indicates that there is some information that it was not able to learn.

Hyperparameter Tuning Using Bayesian Optimization

So LightGBM is better for now, but we want to tune the two models to be able to see how the tuned versions perform. This could make a big difference.

Rather than using GridSearchCV, we'll be using **Bayesian optimization** here. At this stage, you should be comfortable with the two approaches for hyperparameter tuning that were presented earlier: GridSearchCV and RandomizedSearchCV.

The Theory of Bayesian Optimization

In Bayesian optimization, rather than trying out every combination of hyperparameters or just trying out random guesses, it allows you to make intelligent guesses by modeling the likelihood of a certain value for the hyperparameters to be the optimum. This is done by modeling a probability distribution around the hyperparameters.

The concept is shown in Figure 15-4. The picture shows only one hyperparameter, but it can be projected onto a situation with multiple hyperparameters. Bayesian optimization will start sampling a number of values of the hyperparameter and observe the model performance at those points.

The goal is not to randomly fall on one value that performs well, but rather to estimate the *gray zone around the curve*. Some values of the hyperparameter will give better performance than others. While testing values, Bayesian optimization makes a model of the probability distribution of the optimum being at this location.

In this way, Bayesian optimization is much more **intelligent** than the two tuning methods that you have seen until here. It can make intelligent, probability-based guesses about the points that should be tested next. This makes it much more powerful for optimizing hyperparameters.

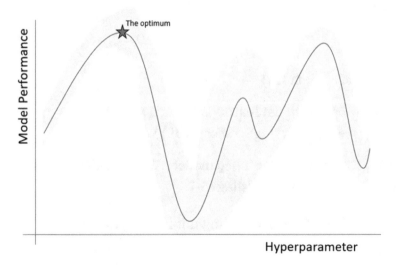

Figure 15-4. Bayesian optimization

Bayesian Optimization Using scikit-optimize

In this part, you will see how to use the scikit-optimize package to apply a Bayesian hyperparameter tuning to XGBoost and LightGBM. After that, you'll do a comparison of the two models. Let's start with XGBoost in Listing 15-6.

Listing 15-6. Applying Bayesian optimization to XGBoost

```
from skopt import BayesSearchCV
from skopt.space import Real, Categorical, Integer
import random
random.seed(0)

xgb_opt = BayesSearchCV(
    XGBRegressor(),
    {
        'learning_rate': (10e-6, 1.0, 'log-uniform'),
        'max_depth': Integer(0, 50, 'uniform'),
        'n_estimators' : (10, 1000, 'log-uniform'),
    },
    n_iter=10,
    cv=3
)

xgb_opt.fit(X_train, y_train)

xgb_tuned_fcst = opt.best_estimator_.predict(X_test)
r2_score(list(y_test), list(xgb_tuned_fcst))
```

The resulting R2 score from the tuned model is **0.969**. Let's continue directly with the LightGBM model, as shown in Listing 15-7.

Listing 15-7. Applying Bayesian optimization to LightGBM

```
Random.seed(0)
lgbm_opt = BayesSearchCV(
    LGBMRegressor(),
    {
        'learning_rate': (10e-6, 1.0, 'log-uniform'),
```

```
        'max_depth': Integer(-1, 50, 'uniform'),
        'n_estimators' : (10, 1000, 'log-uniform'),
    },
    n_iter=10,
    cv=3
)

lgbm_opt.fit(X_train, y_train)

lgbm_tuned_fcst = lgbm_opt.best_estimator_.predict(X_test)
r2_score(list(y_test), list(lgbm_tuned_fcst))
```

The R2 score from LightGBM is **0.973.** So after using Bayesian optimization for the tuning of the two models, the scores have turned around: it is now XGBoost that wins the benchmark. As a final confirmation, let's plot the predictions that have been made by the two models, using the code in Listing 15-8, and you'll obtain the plot in Figure 15-5.

Listing 15-8. Plotting the two tuned models

```
import matplotlib.pyplot as plt
plt.figure(figsize=(20,10))
plt.plot(list(y_test))
plt.plot(list(xgb_tuned_fcst))
plt.plot(list(lgbm_tuned_fcst))
plt.legend(['actual', 'xgb forecast', 'lgbm forecast'])
plt.ylabel('Traffic Volume')
plt.xlabel('Steps in test data')
plt.show()
```

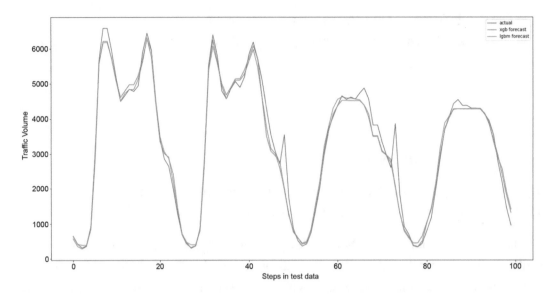

Figure 15-5. *Compare performances of the two tuned models*

In the current execution, **n_iter**, or the number of iterations, was fixed at 20. The execution with this number of iterations is relatively fast. If you want to play around with it, you could test if higher values for n_iter allow the Bayesian optimization to find even higher scores.

Conclusion

In conclusion, you have discovered two very performant machine learning algorithms in this chapter. You have also seen throughout this chapter how you could do a benchmark between two models. In practice, you'll often do benchmarks between even more models. You'll discover more on model benchmarking in the last chapter of this book. Yet it is important to realize that model development should always be data-driven: you choose the model for which you can be confident that it has the best performances in the future.

Key Takeaways

- Gradient boosting is an ensemble learning technique that lets subsequent individual models from the ensemble learn on the parts that are not yet understood well by the model.

- You have seen two models of gradient boosting:

    - *XGBoost*: A more traditional method for gradient boosting

    - *LightGBM*: A newer but very performant competitor

- Bayesian optimization is a more intelligent method for tuning hyperparameters. It estimates the probability of the optimum being on a certain location and therefore makes intelligent guesses for the optimum.

PART V

Advanced Machine and Deep Learning Models

CHAPTER 16

Neural Networks

In the previous five chapters, you have discovered a number of supervised machine learning models, starting from linear regression to gradient boosting. In this chapter, you'll discover **Neural Networks (NNs)**.

The scope of Neural Networks is huge. The version that you'll see in this chapter is a subgroup called **fully connected neural networks**. Those neural networks are intuitively quite close to the previous supervised models. In the following chapters, you'll also discover **Recurrent Neural Networks**, and you'll see a specific network called the **LSTM** Neural Network.

The goal of this chapter is to get familiar with the general idea of Neural Networks.

Fully Connected Neural Networks

So let's start with a schematic overview of the model in Figure 16-1.

© Joos Korstanje 2021
J. Korstanje, *Advanced Forecasting with Python*, https://doi.org/10.1007/978-1-4842-7150-6_16

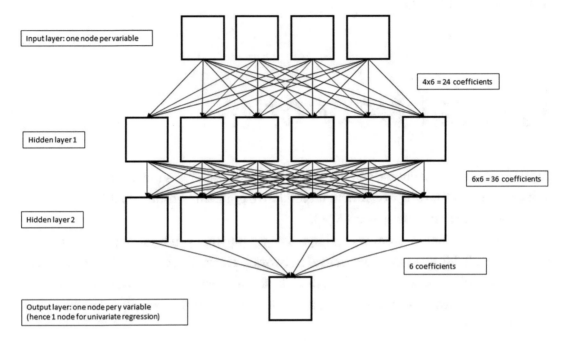

Input layer: one node per variable

4x6 = 24 coefficients

Hidden layer 1

6x6 = 36 coefficients

Hidden layer 2

6 coefficients

Output layer: one node per y variable
(hence 1 node for univariate regression)

Figure 16-1. *Fully connected neural networks schema*

You can read this schema from top to bottom. The model always contains an input layer and an output layer. The input layer of the network contains one node for each variable. The output layer contains the same number of nodes as there are output variables. In the case of univariate regression, there is one output variable (the target variable y), but sometimes there can be multiple output variables at the same time, as you've seen, for example, in the VAR model. In this case, there would be multiple nodes in the output layer.

After the input, you go to the first hidden layer. Roughly speaking, the values of the hidden layer are computed by **multiplying the input by a weight** and then **passing the value through an activation function**. This combination of multiplication and activation functions repeats itself in each node until arriving at the output node. This model is called a fully connected model because each node is connected to each node in the following layer. There are other shapes of architectures in which this is not the case: we'll see some examples in the next two chapters.

Activation Functions

When you choose the architecture of your neural network, you also need to choose the type of activation functions that you want at each location. The activation functions are applied in each node of the network, before going to the next layer.

Three relatively common choices of activation functions are **tanh**, **ReLU**, and **sigmoid**. You can see their specific behaviors in Figure 16-2. The reason that we use activation functions is that they allow the model to fit more complex problems with fewer nodes, as they add nonlinearity into the computation.

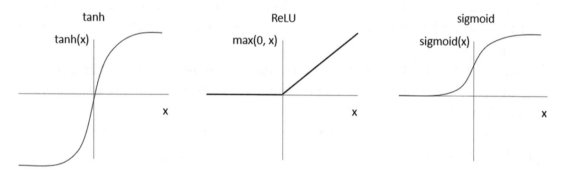

Figure 16-2. *Three common activation layers*

Without getting into too much detail on the activation layers, it is important to know how to use them. ReLU is often a good choice for starting, whereas tanh is often used for Recurrent Neural Networks that you'll see in the next chapter.

The choice for activation layers, just like many things in fitting neural networks, is a choice that is not obvious: trial and error to obtain a working neural network is the only way to go. Sometimes, if you're lucky, you can also find examples in the literature of neural networks that are applied to a similar case as yours. The important thing to remember is that it takes time and effort to come up with something accurate, but if you're successful, you can obtain great performances.

The Weights: Backpropagation

So while moving an input value through your neural network, you encounter something else besides the activation layer: the weights. These weights are estimated at the time of model fitting, through an algorithm called the backpropagation algorithm.

The backpropagation algorithm computes the gradient of the loss function with respect to the weights of the network, one layer at a time. It then iterates backward from the last layer, which is why it is called backpropagation. It is an efficient way for fitting the huge number of weights that is needed for a neural network. Yet it is still a complex algorithm, and it takes time to compute.

Optimizers

The backpropagation algorithm's performances depend on the choice of the optimizer. This is a decision that has to be made again by the modeler. There are many optimizers available. The first available optimizer was gradient descent, as described in the previous paragraph.

SGD, short for **Stochastic Gradient Descent**, is an improved version of the gradient descent algorithm. It is more efficient as it computes not on the whole dataset at every iteration, but only a subset of the dataset.

Two improvements have been made on SGD, and those have delivered two new optimizers: **RMSProp** and **AdaGrad**. An even more improved optimizer called Adam uses a combination of those two.

Although I won't go into the deep mathematical difference between optimizers, it is important to know that there are different optimizers available. Also, a lot of work is still ongoing, and newer and better versions may well arise in the near future. The important thing to retain here is that the choice for an optimizer is a hyperparameter to be chosen by the model developer. Playing around with different optimizers may just help you to add those last points of accuracy to your model.

Learning Rate of the Optimizer

The next thing that you need to choose as a hyperparameter is the learning rate of your optimizer. You can see it as follows. Your optimizer is going to find the right way to move in, and it is going to take a step in that direction. Yet those steps can be either large or small steps. This depends on the learning rate that you choose.

Choosing large learning rates means that you take large steps in the right direction. Yet a risk is that you take too large steps, and therefore you step over the optimum and then miss it.

A small learning rate on the other hand may let you get stuck into a local optimum and not be able to get out, as your step size is too small. You may also take a long time to converge.

Hyperparameters at Play in Developing a NN

Now until here, the overview is not much different than what we've seen before, for example, in Random Forests and XGBoost. Yet there is a big difference in developing Neural Networks: the number of hyperparameters and the time of training the model are so huge that it becomes impossible to simply launch a hyperparameter optimization tool.

This is all due to the backpropagation algorithm used for fitting the neural network. In short, this algorithm goes back and forth through the network, and it updates the weights. It does not pass all the data at once, but it passes batch by batch until all the data has been passed. When all data has been passed, this is called an **epoch**.

As you have seen, for neural networks, a lot of hyperparameter tuning has to be done. A lot of this is done by hand, using specific tools to judge the quality of the model fit. Building Neural Networks is much more complex than fitting classical machine learning models.

As an overview, the main neural network's hyperparameters are

- The number of layers.

- The number of nodes in each of the layers.

- The **optimizer** is the method to update the weights throughout backpropagation. One of the standard optimizers is Adam, but there are many more.

- The **learning rate** of the optimizer influences the step size of the optimizer. Too small learning rates make the steps toward the optimum too small and can make it too slow, or you can get blocked in a local optimum.

Two more hyperparameters that you need to choose are less related to the mathematics, but they are still very important:

- The **batch size** specifies the number of individuals that will be used for each pass through the algorithm. If it's too large, you may run out of RAM; but if it's too small, it may be too slow.

- The number of **epochs** is the number of times that the whole dataset is passed through the network. The more epochs, the longer the model continues training. But this should really depend on at which moment you reach the optimum.

Introducing the Example Data

Before moving on to the fitting and optimization of the neural network, let's introduce the example data and two data preparation methods. In this chapter, we'll be using the example data from the Max Planck Institute for Biogeochemistry. The dataset is called the Jena Climate dataset. It contains weather measures like temperature, humidity, and more, recorded every 10 minutes.

You can download the data from https://storage.googleapis.com/tensorflow/ tf-keras-datasets/jena_climate_2009_2016.csv.zip. You can also download and unzip it and use pandas to import it with Listing 16-1.

Listing 16-1. Importing the data

```
import keras
import pandas as pd
from zipfile import ZipFile
import os

uri = "https://storage.googleapis.com/tensorflow/tf-keras-datasets/jena_
climate_2009_2016.csv.zip"
zip_path = keras.utils.get_file(origin=uri, fname="jena_climate_2009_2016.
csv.zip")
zip_file = ZipFile(zip_path)
zip_file.extractall()
csv_path = "jena_climate_2009_2016.csv"

df = pd.read_csv(csv_path)
del zip_file
```

```
df = df.drop('Date Time', axis=1)
cols = ['p',  'T', 'Tpot', 'Tdew', 'rh', 'VPmax', 'VPact', 'VPdef', 'sh',
'H2OC', 'rho', 'wv', 'mwv', 'wd']
df.columns = cols
```

In this example, we'll do a **forecast of the temperature 12 hours later**. To do this, we create lagged variables for the independent variables and make a correct dataframe. We have quite a lot of data, so we can add multiple lagged values to add the most information possible to the model. We'll use Listing 16-2 to add the values for 72 lags (72 times 10 minutes, and then for 84, 96, 108, 120, and 132 steps back in time).

Listing 16-2. Creating the lagged dataset

```
y = df.loc[2*72:,'T']
lagged_x = []
for lag in range(72,2*72,12):
  lagged = df.shift(lag)
  lagged.columns = [x + '.lag' + str(lag) for x in lagged.columns]
  lagged_x.append(lagged)

df = pd.concat(lagged_x, axis=1)
df = df.iloc[2*72:,:] #drop missing values due to lags
```

Specific Data Prep Needs for a NN

Neural Networks are very sensitive to problems with the input data. Let's have a look at two tools that are often useful in data preparation.

Scaling and Standardization

A neural network will not be able to learn if you do not standardize the input data. Standardizing means *getting the data onto the same scale*. Two examples for this are a standard scaler and a MinMax scaler:

1. A standard scaler maps a variable to follow a standard normal distribution. That is, the new mean of the variable is 0, and the

new standard deviation is 1. It is obtained by taking each value minus the average of the variable and then dividing it through the standard deviation.

2. The MinMax scaler brings a variable into the range of 0–1 by subtracting the variable's minimum from each value and then dividing it by the range of the variable.

You can apply a scaler using the syntax in Listing 16-3.

Listing 16-3. Fitting the MinMaxScaler

```
# apply a min max scaler
from sklearn.preprocessing import MinMaxScaler
scaler = MinMaxScaler()
df = pd.DataFrame(scaler.fit_transform(df), columns = df.columns)
```

Principal Component Analysis (PCA)

The **PCA** is a machine learning model in itself. It comes from the family of dimension reduction models. It allows to take a dataset with a large number of variables and *reduce the number of variables into a projection onto a number of dimensions*. Those dimensions will contain the larger part of the information and will contain much less noise.

Making sure that the input data contains less noise will strongly help during the fitting of your Neural Network. So how does the PCA work? The idea is to make new variables, called **components**, based on combinations of strongly correlated variables. In Figure 16-3, you can see a hypothetical example with two variables Rain and Humidity that you can expect to be correlated. The first principal component captures the most possible variation.

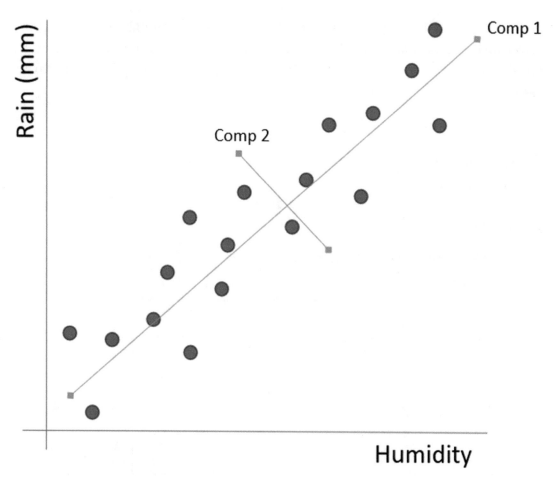

Figure 16-3. *PCA*

The component is a mathematical formula that is a linear combination of the original variables. You can use this score as a new variable. If the principal component is capturing a lot of the original variables, it can be interesting to use the component in your machine learning model rather than the original variables.

To fit a PCA, you generally start with a PCA with all the components. This is done in Listing 16-4.

Listing 16-4. Fitting the full PCA

```
# Fit a PCA with maximum number of components
from sklearn.decomposition import PCA
mypca = PCA()
mypca.fit(df)
```

You use this PCA to make a **scree plot**. A scree plot is one of multiple tools used to decide on the number of components to retain. At the point of the elbow, you choose the number of components. You can use Listing 16-5 to make a scree plot. It is shown in Figure 16-4.

Listing 16-5. Making a scree plot

```
# Make a scree plot
import matplotlib.pyplot as plt
plt.plot(mypca.explained_variance_ratio_)
plt.show()
```

You can see the scree plot in Figure 16-4.

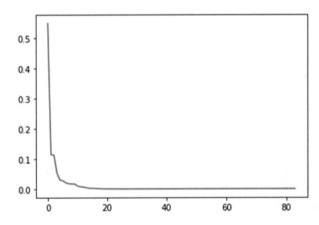

Figure 16-4. *Scree plot*

In this plot, the x-axis shows the components from the first component to the last. The y variable shows the amount of variation that is captured in those components. So you clearly see that the first *five to ten components* have much more information in them than the higher components (those more to the right). You could see the elbow being somewhere around 5, but it seems a good idea to retain ten components so that we retain almost all information but while having ten variables rather than over 80. Finally, you refit the PCA with ten components and transform the data using Listing 16-6.

Listing 16-6. Fitting the PCA with ten components

```
mypca = PCA(10)
df = mypca.fit_transform(df)
```

The Neural Network Using Keras

Now that we have made sure that our data are correctly prepared, we can finally move on to the actual neural network.

Building Neural Networks is a lot of work, and I want to find a good balance in showing you the way to get started and to work on improving your network rather than just showing a final performant network.

A general great first start is to start with a relatively simple network and work your way up from there. In this case, let's start with a network using two dense layers with 64 nodes. That would make the architecture look as follows.

For the other hyperparameters, let's take things that are a little bit standard:

- *Optimizer*: Adam

- *Learning rate*: 0.01

- *Batch size*: 32 (reduce this if you don't have enough RAM)

- *Epochs*: 10

Before starting, let's do a train-test split as shown in Listing 16-7.

Listing 16-7. Train-test split

```
from sklearn.model_selection import train_test_split

X_train, X_test, y_train, y_test = train_test_split(df, y, test_size=0.33,
random_state=42)
```

Now you build the model using the **keras library** using the following code. First, you specify the architecture using Listing 16-8. Keras is the go-to library for neural networks in Python.

Listing 16-8. Specify the model and its architecture

```
from tensorflow.keras.models import Sequential
from tensorflow.keras.layers import Dense
import random
random.seed(42)

simple_model = Sequential([
  Dense(64, activation='relu', input_shape=(X_train.shape[1],)),
  Dense(64, activation='relu'),
  Dense(1),
])
```

You can obtain a summary to check if everything is alright using Listing 16-9.

Listing 16-9. Obtain a summary of the model architecture

```
simple_model.summary()
```

Then you compile the model using Listing 16-10. In the compilation part, you specify the optimizer and the learning rate. You also specify the loss, in our case the Mean Absolute Error.

Listing 16-10. Compile the model

```
simple_model.compile(
  optimizer=keras.optimizers.Adam(learning_rate=0.01),
  loss='mean_absolute_error',
  metrics=['mean_absolute_error'],
)
```

And then you fit the model using Listing 16-11. At the fitting call, you specify the epochs and the batch size. You can also specify a validation split so that you obtain a train-validation-test scenario in which you still have the test set for a final check of the R2 score that is not biased by the model development process.

Listing 16-11. Fit the model

```
smod_history = simple_model.fit(X_train, y_train,
        validation_split=0.2,
        epochs=10,
        batch_size=32,
        shuffle = True
)
```

Be aware that fitting neural networks can take a lot of time. Running on GPU is generally fast but not always possible depending on your computer hardware.

Now the important part here is to figure out whether or not this model has learned something using those hyperparameters. There is a key graph that is going to help you infinitely while building neural networks. You can obtain this graph using Listing 16-12 and see it in Figure 16-5.

Be aware that you may get slightly different results. Setting the random seed is not enough to force randomness to be the same in Keras. Although it is possible to force exact reproducibility in Keras, it is quite complex, so I prefer to leave it out and accept that results are not 100% reproducible. You can check out these instructions for more information on how to fix the randomness in Keras: https://keras.io/getting_ started/faq/#how-can-i-obtain-reproducible-results-using-keras-during- development.

Listing 16-12. Plot the training history

```
plt.plot(smod_history.history['loss'])
plt.plot(smod_history.history['val_loss'])
plt.title('model loss')
plt.ylabel('accuracy')
plt.xlabel('epoch')
plt.legend(['train', 'val'], loc='upper left')
plt.show()
```

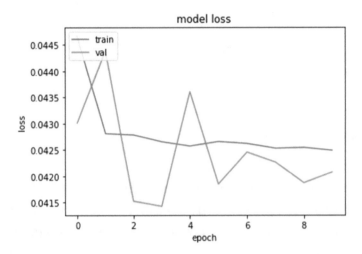

Figure 16-5. *History plot that shows a model that doesn't learn too well*

In Figure 16-6, you see a few examples of graphs that you are or are not looking for. In graph 1, you see the ideal curve that you are trying to obtain.

In graph 2, you see a neural net that is learning well, and it also shows overfit that happens in practice: the training loss goes down a lot, and the validation loss does too. Yet at some point, you need to reduce the number of epochs, or else your model will start to overfit. If you remember from a previous chapter, overfitting means that the model is learning on specifics from the training data that are actually noise and therefore the learned trends do not generalize.

In the third graph, you see a neural net that is not learning anything. You can see this because the training loss is not going down. In this case, you may either have a problem with your data, or you might want to test out a very different network setup.

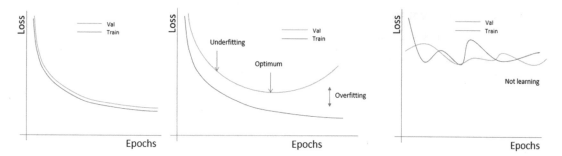

Figure 16-6. *Ideal history plots*

Now in our case, what happened? Most likely a case 3: the model is not learning anything. And why is the model not learning? We should try out whether there are just too few neurons to learn it. From here on, you know that you need to increase the number of layers and neurons and see whether this is getting any better.

From here on, it is honestly a real process of trial and error. The first model that you should try is generally a very simple one, and it will be not learning enough. The next model that you should search for is a model that does learn, even if in the worst case it is overfitting. Then as the last step, you fine-tune until you obtain a graph close enough to the graph on the left in Figure 16-6 and you obtain an error score that is good enough to you.

After trial and error, a better model architecture that has been found is the one in Listing 16-13. Far from saying that this is the best model, at least this model is able to obtain a relatively good R2 score on the test data of **0.908**. Feel free to try and tweak this model more and see what happens and whether you can improve it.

Listing 16-13. A better architecture

```
random.seed(42)
model = Sequential([
  Dense(256, activation='relu', input_shape=(X_train.shape[1],)),
  Dense(256, activation='relu'),
  Dense(256, activation='relu'),
  Dense(256, activation='relu'),
  Dense(256, activation='relu'),
  Dense(256, activation='relu'),
  Dense(256, activation='relu'),
  Dense(256, activation='relu'),
  Dense(256, activation='relu'),
```

```
  Dense(256, activation='relu'),
  Dense(256, activation='relu'),
  Dense(256, activation='relu'),
  Dense(256, activation='relu'),
  Dense(256, activation='relu'),
  Dense(256, activation='relu'),
  Dense(256, activation='relu'),
  Dense(256, activation='relu'),
  Dense(256, activation='relu'),
  Dense(256, activation='relu'),
  Dense(256, activation='relu'),
  Dense(1), ])
model.compile(
  optimizer=keras.optimizers.Adam(learning_rate=0.001),
  loss='mean_absolute_error',
  metrics=['mean_absolute_error'],
)
history = model.fit(X_train, y_train,
           #validation_data=(X_test, y_test),
           validation_split=0.2,
           epochs=100,
           batch_size=32,
           shuffle = True
)
plt.plot(history.history['loss'])
plt.plot(history.history['val_loss'])
plt.title('model loss')
plt.ylabel('accuracy')
plt.xlabel('epoch')
plt.legend(['train', 'val'], loc='upper left')
plt.show()

preds = model.predict(X_test)
print(r2_score(preds, y_test))
```

The resulting graph is shown in Figure 16-7.

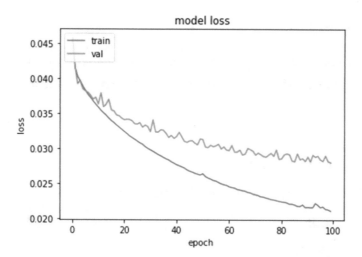

Figure 16-7. *History plot of the more complex model*

Conclusion

The really difficult thing with fitting neural networks is that it is very hard to know if you've achieved something that could be improved or whether it's quite good already. For now, this chapter has shown how to work on improving from a very simple network and trying to get into an optimized version.

There are some libraries that you could use for hyperparameter optimization. Of course, you could even grid search your architecture. Just know that there is no wonder cure for fitting neural networks. It requires much more effort and dedication than classic machine learning models. Training times are very long, and you will need significant computing power to start thinking about such optimization. Yet on the upside, neural networks can sometimes obtain results that are much more powerful than classical machine learning, and it is an important tool to have in your machine learning toolbox.

There are a lot of developments in the scientific field of neural networks and AI. In the coming two chapters, you'll discover neural network structures that are more advanced than the dense structure, but that may apply very well in the case of forecasting.

Key Takeaways

- Neural Networks are a powerful supervised machine learning model, but it is more difficult to build it than classical machine learning models.

- The hyperparameters are

    - The number of layers

    - The number of nodes in each of the layers

    - The **optimizer**

    - The **learning rate**

    - The **batch size**

    - The number of **epochs**

- The graph showing the descent of the train and validation loss is a very important aspect in developing neural networks.

- PCA is a model for dimension reduction. It can be also be used for data preparation, especially when there are many correlated variables.

- Scaling the input data is necessary for neural networks. Some commonly used methods are the standard scaler and the MinMax scaler.

CHAPTER 17

RNNs Using SimpleRNN and GRU

In this chapter, you'll discover a more advanced version of neural networks called **Recurrent Neural Networks**. There are three very common versions of RNNs: SimpleRNN, GRU (Gated Recurrent Unit), and LSTM (Long Short Term Memory). In practice, SimpleRNNs are hardly used anymore for a number of problems that I'll talk about later. Therefore, I have regrouped SimpleRNN and GRU in this chapter, and LSTMs have their own chapter.

What Are RNNs: Architecture

So what are Recurrent Neural Networks, and what makes them different from the dense neural networks that you've seen just before? The following graph shows a node of a fully connected net against an example with a SimpleRNN node.

© Joos Korstanje 2021
J. Korstanje, *Advanced Forecasting with Python*, https://doi.org/10.1007/978-1-4842-7150-6_17

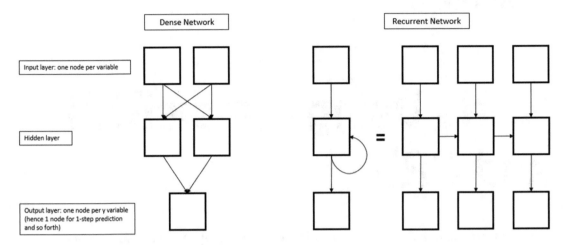

Figure 17-1. *Recurrent network vs. dense network*

As you can see in Figure 17-1, the big difference in the RNN block is that there is a feedback loop. Where each input of a fully connected network is completely independent, the inputs of an RNN have a feedback relation with each other. This makes **RNNs great for data that has sequence**s, for example:

- Time series

- Written text (sequences of words)

- DNA sequences

- uGeolocation sequences

Inside the SimpleRNN Unit

Due to this difference in architecture, there is also a difference inside the units. As you can see, there is not just one input in each unit, but there are two inputs. Those two inputs have to be taken into account. As a schematic overview, this looks as shown in Figure 17-2, in which X is the input at time t, Y is the target variable at time t, and the a's are the weights that go from left to right in Figure 17-1, so the weights that make the model fit as a sequence through time.

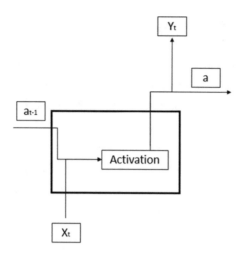

Figure 17-2. *RNN cell*

In the previous chapter, you have mainly seen the **ReLU** activation layer being used. In Recurrent Neural Networks, however, the **tanh** layer is the standard. The reason for this is that for long sequences, the ReLU layer suffers from the exploding gradient problem: the repeated multiplication with weights is acting like an exponentiation that makes it explode. The tanh activation layer does not have this problem, as the values are forced to stay between -1 and 1.

The Example

The RNN learns on sequences. Therefore, we will have to adapt the problem statement. In this chapter, let's work with the same data as in the previous chapter to get a good feel of how the data preparation and use of the data are different from fully connected architectures.

Just to remember quickly: You have a dataset with measures on weather data, and we try to predict the temperature 12 hours (72 time steps of 10 minutes) into the future.

Now what we did in the fully connected model was to create lagged variables. There was one y variable (which was not lagged), and the independent variables were lagged values of the y variable and a lot of other variables. The first lag was at 72 time steps, so that the model would use the data from 12 hours ago to predict the now (having a history from now to 12 hours back to predict the now is equivalent to having data from now to predict 12 hours later).

Predicting a Sequence Rather Than a Value

In the RNN, this is not what we should do, as the RNN learns sequences. A big jump of 72 time steps is not really respecting the sequential variation. Yet we do not just want to predict one time step later, as that would mean that we predict the temperature in 10 minutes from now, not really interesting.

What we will do is *create a matrix of y variables*, with lags as well. Before, we wanted to predict one value 72 time steps into the future, but let's model to predict each of those 72. Even though it might be possible to do it with one y variable, this is also an interesting case of *multistep forecasting*, which is often useful.

Univariate Model Rather Than Multivariable

Another thing that we change from the previous model is that in this case, we will use only the temperature data and not the other variables. This may make the task slightly harder to accomplish, but it'll be easier to get your head around the use of sequences and RNNs. However, you must know that it is possible to add other explanatory variables into an RNN. For forecasting tomorrow's temperature, you may want to use not only today's temperature but also today's wind direction, wind speed, and humidity, for example. In this case, you could add a third dimension to the input data.

Preparing the Data

You can prepare the data using the following steps. The first step is to import the data using Listing 17-1.

Listing 17-1. Importing the data

```
import keras
import pandas as pd

from zipfile import ZipFile
import os

uri = "https://storage.googleapis.com/tensorflow/tf-keras-datasets/
jena_climate_2009_2016.csv.zip"
```

```
zip_path = keras.utils.get_file(origin=uri, fname="jena_climate_2009_2016.
csv.zip")
zip_file = ZipFile(zip_path)
zip_file.extractall()
csv_path = "jena_climate_2009_2016.csv"

df = pd.read_csv(csv_path)
del zip_file
```

The next step is to delete all columns other than the temperature, as we are building a univariate model. This is done in Listing 17-2.

Listing 17-2. Keep only temperature data

```
df = df[['T (degC)']]
```

Now, as you remember from Chapter 16, neural networks need data that has been standardized. So let's apply a MinMaxScaler as in Listing 17-3.

Listing 17-3. Apply a MinMaxScaler

```
# apply a min max scaler
from sklearn.preprocessing import MinMaxScaler
scaler = MinMaxScaler()
df = pd.DataFrame(scaler.fit_transform(df), columns = ['T'])
```

Now a part that may be harder to get your head around intuitively. We need to split the data into a shape in which we have sequences of past data and sequences of future data. We want to predict 72 steps into the future, and we'll use 3*72 steps into the past. This is an arbitrary choice, and please feel free to try out using more or less past data. The code in Listing 17-4 loops through the data and creates sequences for the model training.

Listing 17-4. Preparing the sequence data

```
ylist = list(df['T'])

n_future = 72
n_past = 3*72
total_period = 4*72
```

```
idx_end = len(ylist)
idx_start = idx_end - total_period

X_new = []
y_new = []
while idx_start > 0:
  x_line = ylist[idx_start:idx_start+n_past]
  y_line = ylist[idx_start+n_past:idx_start+total_period]

  X_new.append(x_line)
  y_new.append(y_line)

  idx_start = idx_start - 1

# converting list of lists to numpy array
import numpy as np
X_new = np.array(X_new)
y_new = np.array(y_new)
```

Now that we have obtained an X and a Y matrix for the model training, as always we need to split it into a train and test set in order to be able to do a fair model evaluation. This is done in Listing 17-5.

Listing 17-5. Splitting into train and test

```
from sklearn.model_selection import train_test_split

X_train, X_test, y_train, y_test = train_test_split(X_new, y_new,
test_size=0.33, random_state=42)
```

A final step is necessary for fitting the SimpleRNN. This is a step that may be hard to understand intuitively, as there is not really a reason for it. The SimpleRNN layer needs an input format that is 3D, and the shape has to correspond to (n_samples, n_timesteps, n_features). This can be obtained using reshape. This reshape is, unfortunately, necessary sometimes when using Keras, as it is very specific as to the exact shape of the input data.

When working with layer types that you don't know yet, this can sometimes give real headaches. But it is crucial to learn how to work with it. Listing 17-6 shows you how to reshape the data into the right format for Keras to recognize it.

Listing 17-6. Reshape the data to be recognized by Keras

```
batch_size = 32

n_samples = X_train.shape[0]
n_timesteps = X_train.shape[1]
n_steps = y_train.shape[1]
n_features = 1

X_train_rs = X_train.reshape(n_samples, n_timesteps, n_features )

X_test_rs = X_test.reshape(X_test.shape[0], n_timesteps, n_features )
```

A Simple SimpleRNN

Now let's start with a very small neural network using the SimpleRNN layer. You can parameterize it with Listing 17-7.

Listing 17-7. Parameterize a small network with SimpleRNN

```
import random
random.seed(42)

from tensorflow.keras.models import Sequential
from tensorflow.keras.layers import Dense, SimpleRNN

simple_model = Sequential([
  SimpleRNN(8, activation='tanh',input_shape=(n_timesteps, n_features)),
  Dense(y_train.shape[1]),
])

simple_model.summary()

simple_model.compile(
  optimizer=keras.optimizers.Adam(learning_rate=0.001),
  loss='mean_absolute_error',
  metrics=['mean_absolute_error'],
)
```

```
smod_history = simple_model.fit(X_train_rs, y_train,
          validation_split=0.2,
          epochs=5,
          batch_size=batch_size,
          shuffle = True
)

preds = simple_model.predict(X_test_rs)

from sklearn.metrics import r2_score
print(r2_score(preds, y_test))

import matplotlib.pyplot as plt
plt.plot(smod_history.history['loss'])
plt.plot(smod_history.history['val_loss'])
plt.title('model loss')
plt.xlabel('epoch')
plt.legend(['train', 'val'], loc='upper left')
plt.show()
```

The obtained plot is shown in Figure 17-3, and the obtained R2 score is **0.66**. This is not a great score, and the training history confirms that the model is not correctly specified: it is going down, but it seems to be not learning enough. The loss is what we call the accuracy throughout the model, as the error function of the neural network is often referred to as the loss function. A more complex model should allow to let the loss go down more.

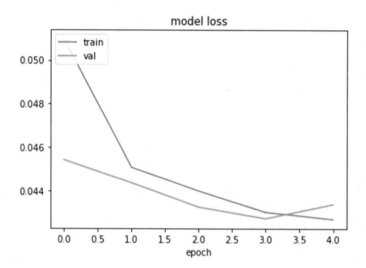

Figure 17-3. *Training history of the SimpleRNN*

SimpleRNN with Hidden Layers

As an improvement to this, let's add a second layer to this network. To do this, you need to specify in every layer but the last '**return_sequences = True**'. Fitting RNNs is even slower than fitting **feedforward neural networks**, so be aware that the computation times in this chapter may be long.

You could build the network in Listing 17-8 that is slightly more complex and see how it performs.

Listing 17-8. A more complex network with three layers of SimpleRNN

```
random.seed(42)

simple_model = Sequential([
  SimpleRNN(32, activation='tanh',input_shape=(n_timesteps, n_features),
  return_sequences=True),
  SimpleRNN(32, activation='tanh', return_sequences = True),
  SimpleRNN(32, activation='tanh'),
  Dense(y_train.shape[1]),
])

simple_model.summary()
```

```
simple_model.compile(
  optimizer=keras.optimizers.Adam(learning_rate=0.001),
  loss='mean_absolute_error',
  metrics=['mean_absolute_error'],
)

smod_history = simple_model.fit(X_train_rs, y_train,
          validation_split=0.2,
          epochs=5,
          batch_size=batch_size,
          shuffle = True
)

preds = simple_model.predict(X_test_rs)

print(r2_score(preds, y_test))

plt.plot(smod_history.history['loss'])
plt.plot(smod_history.history['val_loss'])
plt.title('model loss')
plt.xlabel('epoch')
plt.legend(['train', 'val'], loc='upper left')
plt.show()
```

This gives a multivariate R2 score of 0.90. You will observe the history plot shown in Figure 17-4.

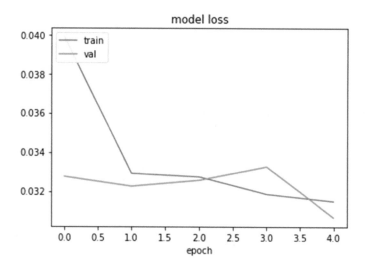

Figure 17-4. *Training history of the three-layer SimpleRNN*

This network allows the loss to go down earlier and lower, even though the curve is still not looking very impressive. At the same time, the obtained R2 is quite good, so it feels like things are moving in the right direction.

Simple GRU

Now we could go further into the SimpleRNN and try to optimize it. Yet we won't do that here. The SimpleRNN is not up to today's standards anymore. It has some serious problems. One big effect of this is that it is not able to remember a very long time back. There are also some more technical issues with it.

A more advanced RNN layer has been invented called GRU, for Gated Recurrent Unit. The GRU cell has more parameters as is shown in Figure 17-5. This shows that there is an extra passage inside the cell, and this allows for an extra parameter to be estimated. This helps with learning long-term trends.

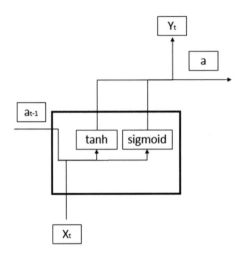

Figure 17-5. *The GRU cell*

Due to those differences, the GRU obtains better general performances than the SimpleRNN, and the SimpleRNN has become very little used. Let's build a simple network with GRU and see how it performs on the data. This is done in Listing 17-9.

Listing 17-9. A simple architecture with one GRU layer

```
random.seed(42)
from tensorflow.keras.layers import GRU

simple_model = Sequential([
   GRU(8, activation='tanh',input_shape=(n_timesteps, n_features)),
  Dense(y_train.shape[1]),
])

simple_model.summary()

simple_model.compile(
  optimizer=keras.optimizers.Adam(learning_rate=0.01),
  loss='mean_absolute_error',
  metrics=['mean_absolute_error'],
)

smod_history = simple_model.fit(X_train_rs, y_train,
         validation_split=0.2,
```

```
        epochs=10,
        batch_size=batch_size,
        shuffle = True
)

preds = simple_model.predict(X_test_rs)

print(r2_score(preds, y_test))

plt.plot(smod_history.history['loss'])
plt.plot(smod_history.history['val_loss'])
plt.title('model loss')
plt.xlabel('epoch')
plt.legend(['train', 'val'], loc='upper left')
plt.show()
```

You will obtain an R2 of **0.88**, which is not too bad for a one-layer model, and you obtain the history plot that is shown in Figure 17-6. The history plot on the other hand shows that the fitting really did not go very well. The loss is not going down as it is supposed to be, and some changes should be made to the model architecture.

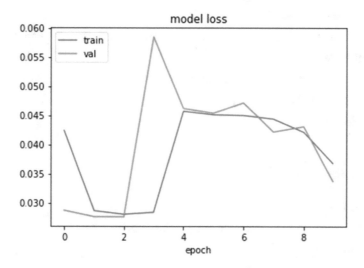

Figure 17-6. *The history of the one-layer GRU model*

GRU with Hidden Layers

Now as a next step, let's try to improve on this network by adding some more layers and see if we can get the loss to go down more. To be totally transparent here, it takes a lot of time to work on the optimization of the architecture and hyperparameters. It is a work of trial and error while sticking to a number of model metrics, including loss and R2 score, but also the loss graph.

It also takes experience to test out different hyperparameters and to develop a feel for the type of changes that you'd need to make. But that's totally normal; there simply is a learning curve in those models. It is also important to realize that it takes a lot of time for one model to fit. This makes it hard to run grid search or other hyperparameter optimizations, yet at the same time, it can make it frustrating to wait a long time only to see a bad score arriving.

The model that you see in Listing 17-10 obtains an R2 score of **0.95**, so quite a lot better than the simple GRU model.

Listing 17-10. A more complex network with three layers of GRU

```
random.seed(42)

simple_model = Sequential([
   GRU(64, activation='tanh',input_shape=(n_timesteps, n_features),
   return_sequences=True),
    GRU(64, activation='tanh', return_sequences=True),
    GRU(64, activation='tanh'),
  Dense(y_train.shape[1]),
])

simple_model.summary()

simple_model.compile(
  optimizer=keras.optimizers.Adam(learning_rate=0.001),
  loss='mean_absolute_error',
  metrics=['mean_absolute_error'],
)

smod_history = simple_model.fit(X_train_rs, y_train,
```

```
        validation_split=0.2,
        epochs=10,
        batch_size=batch_size,
        shuffle = True
)

preds = simple_model.predict(X_test_rs)

print(r2_score(preds, y_test))

plt.plot(smod_history.history['loss'])
plt.plot(smod_history.history['val_loss'])
plt.title('model loss')
plt.xlabel('epoch')
plt.legend(['train', 'val'], loc='upper left')
plt.show()
```

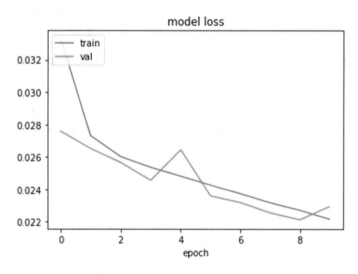

Figure 17-7. *The history of the three-layer GRU model*

The history graph of this model is shown in Figure 17-7. It is probably not the best model, but at least we see that the training and validation losses are both going down. The obtained R2 is not too bad. I leave it as an exercise for you to try and improve on this architecture using the loss graph and model metrics.

Key Takeaways

- Recurrent Neural Networks have an integrated feedback loop that makes them great for modeling sequences.

- There are multiple types of RNN cells:

  - The SimpleRNN is a basic RNN cell that is not much used anymore as it has numerous theoretical problems and is not able to learn long-term trends.

  - The GRU cell is an improvement of the RNN cell, and it has additional parameters that make it easier to remember long-term trends.

  - The LSTM will be covered in Chapter 18.

- RNNs, especially when training on long sequences, benefit from using a tanh activation function rather than the ReLU that is common in dense networks.

- You can use RNNs to predict a sequence of multiple steps to make a multistep forecast.

- You have seen a number of practical modeling examples, and this will help you to understand which history plots are good or not.

CHAPTER 18

LSTM RNNs

In the previous chapter, you have discovered two types of Recurrent Neural Network cells called SimpleRNN and GRU (Gated Recurrent Unit). In this chapter, you'll discover a third type of cell called **LSTM,** for **Long Short-Term Memory**.

What Is LSTM

LSTMs as a third and last type of RNN cell are even more advanced than the GRU cell. If you remember from the last chapter, the SimpleRNN cell allows having recurrent architectures, by adding a feedback loop between consecutive values. It therefore is an improvement on "simple" feedforward cells that do not allow for this.

A problem with the SimpleRNN is the longer-term trends, which you could call a longer-term memory. The GRU cell is an improvement on the SimpleRNN that adds a weight to the cell that serves to learn longer-term processes.

The LSTM cell adds long-term memory in an even more performant way because it allows even more parameters to be learned. This makes it the most powerful RNN to do forecasting, especially when you have a longer-term trend in your data. LSTMs are one of the state-of-the-art models for forecasting at this moment.

The LSTM Cell

Let's see how the LSTM cell differs from the GRU cell. It has much more components. Notably, there are *three sigmoids and two tanh operations* all combined into the same cell. You can see a schematic overview in Figure 18-1 where there are two weights coming in from the past cell (c and a of the time t-1) and once transformed another two weights are going (c and a of the time t). The X is the input to this cell, and the Y is the output.

© Joos Korstanje 2021
J. Korstanje, *Advanced Forecasting with Python*, https://doi.org/10.1007/978-1-4842-7150-6_18

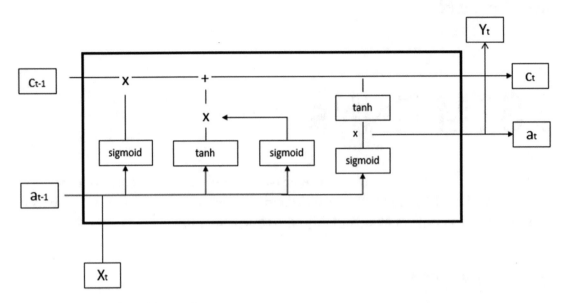

Figure 18-1. *The LSTM cell*

In practice, this all comes down to having more parameters to estimate, and the LSTM can therefore better fit on data that have short- and long-term trends.

Example

In applied machine learning, it is important to benchmark models against each other. Let's therefore use the same data for a last time and make it into a benchmark of the three RNN models that we have seen. As a short recap, the performances that were obtained by the models in Chapter 17 are repeated in Table 18-1.

Table 18-1. *Performances of the SimpleRNN and GRU Models*

Model	SimpleRNN	GRU
One layer of 8	0.66	0.88
Three layers of 64	0.90	0.95

Let's see how the LSTM compares to this. You can use the same data preparation as used in Chapter 17, which is repeated in Listing 18-1.

Listing 18-1. Importing the weather data

```
import keras
import pandas as pd

from zipfile import ZipFile
import os

uri = "https://storage.googleapis.com/tensorflow/tf-keras-datasets/jena_
climate_2009_2016.csv.zip"
zip_path = keras.utils.get_file(origin=uri, fname="jena_climate_2009_2016.
csv.zip")
zip_file = ZipFile(zip_path)
zip_file.extractall()
csv_path = "jena_climate_2009_2016.csv"

df = pd.read_csv(csv_path)
del zip_file

# retain only temperature
df = df[['T (degC)']]

# apply a min max scaler
from sklearn.preprocessing import MinMaxScaler
scaler = MinMaxScaler()
df = pd.DataFrame(scaler.fit_transform(df), columns = ['T'])

# convert to windowed data sets
ylist = list(df['T'])

n_future = 72
n_past = 3*72
total_period = 4*72

idx_end = len(ylist)
idx_start = idx_end - total_period
```

```
X_new = []
y_new = []
while idx_start > 0:
    x_line = ylist[idx_start:idx_start+n_past]
    y_line = ylist[idx_start+n_past:idx_start+total_period]

    X_new.append(x_line)
    y_new.append(y_line)

    idx_start = idx_start - 1

import numpy as np
X_new = np.array(X_new)
y_new = np.array(y_new)

# train test split
from sklearn.model_selection import train_test_split
X_train, X_test, y_train, y_test = train_test_split(X_new, y_new,
test_size=0.33, random_state=42)

# reshape data into the right format for RNNs
n_samples = X_train.shape[0]
n_timesteps = X_train.shape[1]
n_steps = y_train.shape[1]
n_features = 1

X_train_rs = X_train.reshape(n_samples, n_timesteps, n_features )
X_test_rs = X_test.reshape(X_test.shape[0], n_timesteps, n_features )
```

LSTM with One Layer of 8

Now, as we did in the previous chapter, let's start with fitting a simple one-layer network. The code is almost the same as Chapter 17's code; you just need to change the name of the layer into LSTM. The full code is shown in Listing 18-2.

Listing 18-2. One-layer LSTM

```python
import random
from tensorflow.keras.models import Sequential
from tensorflow.keras.layers import Dense, LSTM

random.seed(42)

batch_size = 32

simple_model = Sequential([
    LSTM(8, activation='tanh',input_shape=(n_timesteps, n_features)),
    Dense(y_train.shape[1]),
])

simple_model.summary()

simple_model.compile(
    optimizer=keras.optimizers.Adam(learning_rate=0.01),
    loss='mean_absolute_error',
    metrics=['mean_absolute_error'],
)

smod_history = simple_model.fit(X_train_rs, y_train,
        validation_split=0.2,
        epochs=5,
        batch_size=batch_size,
        shuffle = True
)

preds = simple_model.predict(X_test_rs)

print(r2_score(preds, y_test))

plt.plot(smod_history.history['loss'])
plt.plot(smod_history.history['val_loss'])
plt.title('model loss')
plt.xlabel('epoch')
plt.legend(['train', 'val'], loc='upper left')
plt.show()
```

The results of this model are the following. The R2 score obtained by this model is **0.93**. The plot of the training loss is shown in Figure 18-2.

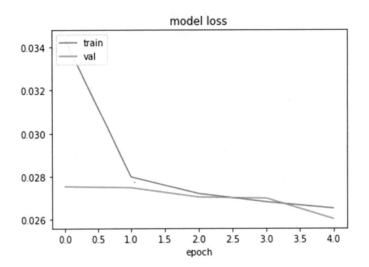

Figure 18-2. *Training history of the one-layer LSTM*

In this graph, you can see that the train loss and val loss both go down a little bit. Yet the drop in validation loss is not very steep. This leads to the conclusion that the model is learning a bit, but that we are probably able to get more out of this model. We'd want to see a bigger drop in validation loss before it plateaus. In the current graph, you're almost directly on a plateau. In Table 18-2, you see the table updated with the new value.

Table 18-2. *Performances of the SimpleRNN and GRU Models*

Model	SimpleRNN	GRU	LSTM
One layer of 8	0.66	0.88	0.93
Three layers of 64	0.90	0.95	

LSTM with Three Layers of 64

As you remember from the previous chapter and as you can see in the table, the GRU model with three layers of 64 cells worked quite well. Before going deeper into the tuning of the LSTM model, let's also try out the performances of this architecture. You can use Listing 18-3 to do this.

Listing 18-3. Three-layer LSTM

```
random.seed(42)

simple_model = Sequential([
    LSTM(64, activation='tanh',input_shape=(n_timesteps, n_features),
    return_sequences=True),
      LSTM(64, activation='tanh', return_sequences=True),
      LSTM(64, activation='tanh'),
  Dense(y_train.shape[1]),
])

simple_model.summary()

simple_model.compile(
  optimizer=keras.optimizers.Adam(learning_rate=0.001),
  loss='mean_absolute_error',
  metrics=['mean_absolute_error'],
)

smod_history = simple_model.fit(X_train_rs, y_train,
          validation_split=0.2,
          epochs=10,
          batch_size=batch_size,
          shuffle = True
)

preds = simple_model.predict(X_test_rs)

print(r2_score(preds, y_test))

plt.plot(smod_history.history['loss'])
plt.plot(smod_history.history['val_loss'])
plt.title('model loss')
plt.ylabel('accuracy')
plt.xlabel('epoch')
plt.legend(['train', 'val'], loc='upper left')
plt.show()
```

This model obtains an R2 score of **0.94**. When comparing this to the other values in Table 18-3, you can observe that it functions a bit less than the GRU three-layer model.

Table 18-3. *Performances of the SimpleRNN and GRU Models*

Model	SimpleRNN	GRU	LSTM
One layer of 8	0.66	0.88	0.93
Three layers of 64	0.90	0.95	**0.94**

Let's also have a look at the history graph shown in Figure 18-3. What we can see in this training graph is that the loss drops more than in the architecture with one layer. The shape of the train loss is looking not too bad: it has the right shape, although it seems that the drop could go on a bit longer, and therefore the plateau would be reached later and at a lower loss. The validation, on the other hand, does not really have the right shape, which tells us that there is probably some R2 to win by tweaking the model.

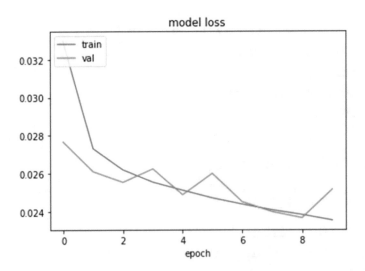

Figure 18-3. *Training history of the three-layer LSTM*

Conclusion

So where does that leave us? The best RNN is the three-layer GRU, very closely followed by the three-layer LSTM and the one-layer LSTM. The differences in scores clearly show why the SimpleRNN layer is not much used anymore. The GRU and the LSTM are two RNN models that are performant for forecasting tasks.

The LSTM model has the advantage over GRU to be able to fit more long-term trends. A hypothetical reason for the better performance of the GRU is that the data did not contain very long-term trends, as the data were reworked to contain only 36 hours of past data for each line of data.

When doing modeling in practice, it is very common to run benchmarks like the one done in this example. The important thing is to master the different models to be able to tune and benchmark them effectively. After that, the decision should be objective based on performances during the benchmark. In this case, it was not the more recent LSTM that won, but the GRU.

Key Takeaways

- The LSTM is the most advanced type of RNN as it has a large number of components that allow it to remember longer-term trends.

- You have seen an example of how to do a model benchmark between multiple models on the same dataset. This is crucial in applied machine learning.

- You have also seen an example of how to interpret and work with neural network history graphs that show the descent in loss. You can use them to draw conclusions on the way your model is learning and how to improve that.

CHAPTER 19

The Prophet Model

In this chapter, you'll discover the **Prophet model**, which was open sourced by Facebook. It is not exactly a model, but rather *an automated procedure* for building forecasting models.

You'll notice that there is a big difference in working with this model than with Keras, as the Prophet model does a lot of the work for you. It has a much higher level of user-friendliness: it does not require any theory to get started, as opposed to the previous chapters.

A disadvantage of this can be that you do have fewer possibilities to understand exactly how and what the model is learning. There are also fewer parameters to tweak, which may be an advantage if the model works, but it may be a disadvantage if you do not obtain the required model precision.

A quote from the developers explains the goal of Facebook's Prophet:

> *We use a simple, modular regression model that often works well with default parameters, and that allows analysts to select the components that are relevant to their forecasting problem and easily make adjustments as needed. The second component is a system for measuring and tracking forecast accuracy, and flagging forecasts that should be checked manually to help analysts make incremental improvements.*

—https://peerj.com/preprints/3190/

As Facebook's Prophet is a procedure rather than an actual model, there is not one particular part of mathematical theory that is contributed by it. It is rather the combination of different processes and the automated way of tuning behind this that are useful. Let's go directly into an example and discover the different components that together make up the Facebook Prophet.

© Joos Korstanje 2021
J. Korstanje, *Advanced Forecasting with Python*, https://doi.org/10.1007/978-1-4842-7150-6_19

The Example

In this example, we'll be working with a dataset from a Kaggle competition on forecasting the number of restaurant visitors. You can download the data on the Kaggle page of the competition that used this dataset: `www.kaggle.com/c/recruit-restaurant-visitor-forecasting`. You will need the following datasets from there: 'air_visit_data.csv.zip', 'date_info.csv.zip', and 'air_reserve.csv.zip'.

We'll be using Prophet, and **Prophet can only make univariate forecasts**. If you remember from earlier, this means that Prophet only makes forecasts with only one dependent variable. If you'd want to adapt to multivariate forecasts, you'd need to train a Prophet for each dependent variable separately. Therefore, we will be working on a combined forecast for the sum of all the restaurants. You can prepare the data of the dependent variable using Listing 19-1.

Listing 19-1. Preparing the dependent variable

```
import pandas as pd
y = pd.read_csv('air_visit_data.csv.zip')
y = y.pivot(index='visit_date', columns='air_store_id')['visitors']
y = y.fillna(0)
y = pd.DataFrame(y.sum(axis=1))
```

The Prophet Data Format

When building a Prophet model, you are always obliged to have a dataset with the columns 'ds' and 'y'. This is how the model recognizes the date column ('ds') and the dependent variable ('y'). You can create this format using Listing 19-2.

Listing 19-2. Preparing the modeling dataframe

```
y = y.reset_index(drop=False)
y.columns = ['ds', 'y']
```

We will also add a train-test split made by putting the last 28 days into the test set. This can be done using Listing 19-3.

Listing 19-3. Creating a train-test split

```
train = y.iloc[:450,:]
test = y.iloc[450:,:]
```

The Basic Prophet Model

Having a dataframe with the two correct column names is all you need to start creating a basic Prophet model. The modeling is not too different from scikit-learn, so it should be quite familiar to you. You can create and fit a basic model using Listing 19-4.

Listing 19-4. Creating a basic Prophet model

```
from fbprophet import Prophet
m = Prophet()
m.fit(train)
```

After the model is fit, you need to make a prediction. To make the prediction, you can use a Prophet function to create a future dataframe that will be an input to the predict function. You can then use the predict method to make the prediction. Listing 19-5 shows how this is done.

Listing 19-5. Making a prediction

```
future = m.make_future_dataframe(periods=len(test))
forecast = m.predict(future)
```

To assess the quality of this forecast, let's compute the R2 score on the test data using Listing 19-6. Once you make the forecast, you obtain a dataframe in which the column **'yhat'** contains the forecast.

Listing 19-6. Computing the R2 score

```
from sklearn.metrics import r2_score
print(r2_score(list(test['y']), list(forecast.loc[450:,'yhat'] )))
```

This basic Prophet model obtains an R2 score of **0.808**. This is already quite good! Let's see how the model is fitting, by plotting the forecast against the test data. The plot is created using Listing 19-7, and it is shown in Figure 19-1.

Listing 19-7. Plotting the fit of the model

```
import matplotlib.pyplot as plt
plt.plot(list(test['y']))
plt.plot(list(forecast.loc[450:,'yhat'] ))
plt.show()
```

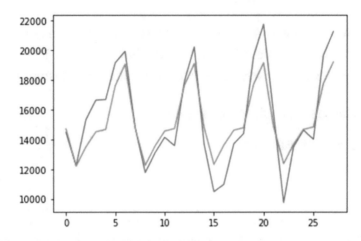

Figure 19-1. *Plot of the prediction against the test data*

There are two other great plots available when using the Prophet model. The first one is showing the forecast against the observed data points for the past and future data. It can be accessed using Listing 19-8. You can see the plot in Figure 19-2.

Listing 19-8. Creating a Prophet forecast plot

```
fig1 = m.plot(forecast)
plt.show()
```

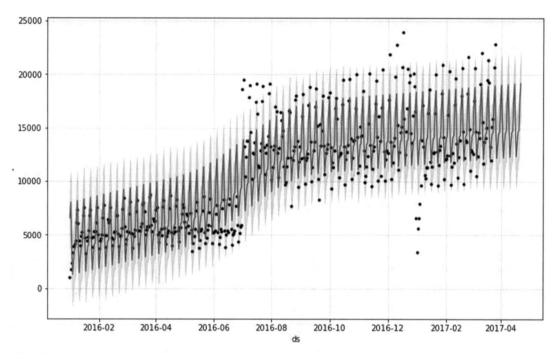

Figure 19-2. *Prophet forecast plot*

The second plot that you can obtain from the Prophet model is a decomposition of the different impacts of the model. This means that the decomposition can show you the impact of the different seasonalities at each time step. This can be done using Listing 19-9. You can see the plot in Listing 19-3.

Listing 19-9. Creating a Prophet decomposition plot

```
fig2 = m.plot_components(forecast)
plt.show()
```

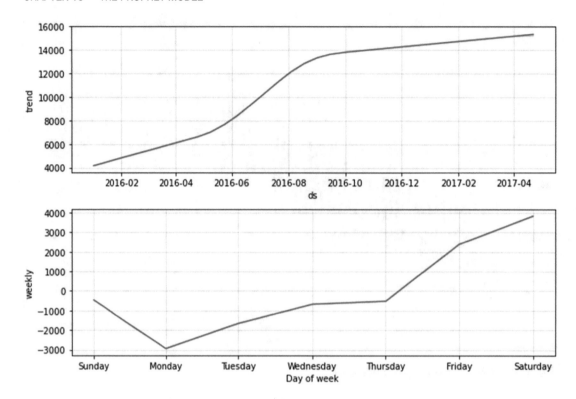

Figure 19-3. *Prophet decomposition plot*

In this plot, you observe the trends that have been fitted. You see that the current model is standard fitting a long-term trend and a weekday effect. The overall trend that the model has fitted seems to be overall increasing, but the growth is slowing down. The weekly plot shows that Saturday is the big day for restaurants and then there is a trend throughout the week.

In the coming parts of this chapter, we'll be adding all possible effects to the model so that the model becomes more and more performant.

Adding Monthly Seasonality to Prophet

As a next step, let's see how to add an additional seasonality effect to Prophet. In this case, we add a monthly seasonality. This is shown in Listing 19-10.

Listing 19-10. Add a monthly seasonality to the plot

```
m2 = Prophet()
m2.add_seasonality(name='monthly', period=30.5, fourier_order=5)

m2.fit(train)

future2 = m2.make_future_dataframe(periods=len(test))
forecast2 = m2.predict(future)
print(r2_score(list(test['y']), list(forecast2.loc[450:,'yhat'] )))

fig2 = m2.plot_components(forecast2)
plt.show()
```

The R2 score of this model is **0.787**. This is slightly lower than the previous model, so we can conclude that adding months has no added value. As this model now has a new effect, you can see that there are now three graphs in the decomposition plot. It is shown in Figure 19-4.

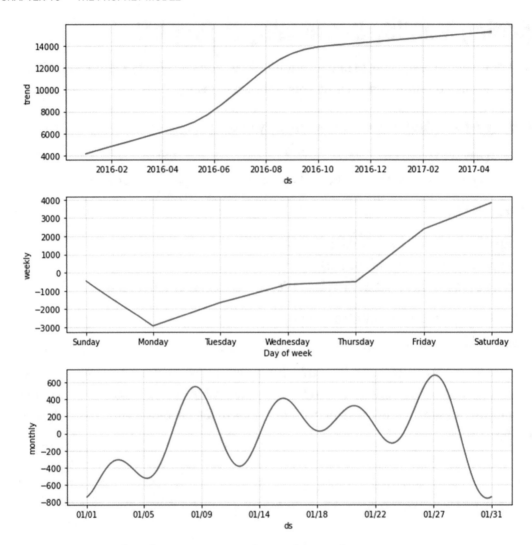

Figure 19-4. *Prophet decomposition plot with months*

Adding Holiday Data to Basic Prophet

As a next step, let's see how to add holiday data to Prophet. Holidays are a standard input in Prophet, which makes it very user-friendly to add holidays. The only thing to do is to make sure that your holiday data are in a dataframe of the right format. The format that you should obtain is a dataframe with

- A column called 'ds' that contains the date of the holiday.

- A column called 'holiday' that contains the name of the holiday.

- Optionally, you can add two columns 'upper_window' and 'lower_window' that contain the number of dates around the holiday that may be impacted by the holiday. For example, Easter and Easter Monday could be modeled together by setting the date on Easter and adding an upper_window of 1.

Listing 19-11 shows how this can be done.

Listing 19-11. Prepare holiday data

```
holidays = pd.read_csv('date_info.csv.zip')
holidays = holidays[holidays['holiday_flg'] == 1]
holidays = holidays[['calendar_date', 'holiday_flg']]
holidays = holidays.drop(['holiday_flg'], axis=1)
holidays['holiday'] = 'holiday'
holidays.columns = ['ds', 'holiday']
```

Then you can fit the model while adding the holiday parameter using Listing 19-12.

Listing 19-12. Add holidays to the model

```
m3 = Prophet(holidays=holidays)
m3.fit(train)
future3 = m3.make_future_dataframe(periods=len(test))
forecast3 = m3.predict(future)

print(r2_score(list(test['y']), list(forecast3.loc[450:,'yhat'] )))

fig2 = m3.plot_components(forecast3)
plt.show()
```

With this model, you obtain an R2 score of **0.806**. It's again not any better than the default model, meaning that the holidays may not help the model to predict better. The plot with the components is shown in Figure 19-5.

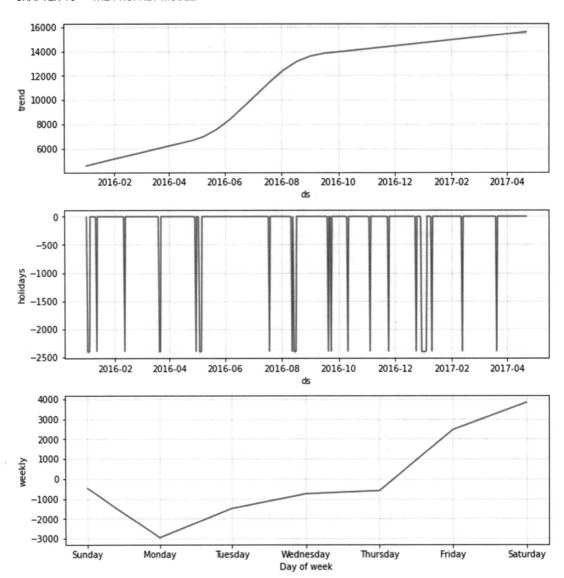

Figure 19-5. *Prophet decomposition plot with holidays*

Adding an Extra Regressor to Prophet

Also available in the dataset is a file with reservations. Of course, we could expect reservations to be a great additional variable to forecast the number of restaurant visitors. You can add an additional regressor to a Prophet model quite easily. First, add the additional variable into the dataframe that contains the columns 'ds' and 'y'. You can do this using Listing 19-13.

Listing 19-13. Add reservations to the data

```
X_reservations = pd.read_csv('air_reserve.csv.zip')
X_reservations['visit_date'] = pd.to_datetime(X_reservations['visit_
datetime']).dt.date
X_reservations = pd.DataFrame(X_reservations.groupby('visit_date')
['reserve_visitors'].sum())
X_reservations = X_reservations.reset_index(drop = False)
train4 = train.copy()
train4['ds'] = pd.to_datetime(train4['ds']).dt.date
train4 = train4.merge(X_reservations, left_on = 'ds', right_on =
'visit_date', how = 'left')[['ds', 'y', 'reserve_visitors']].fillna(0)
```

Then in the model fitting, you can use the method "add_regressor" to add it into the model. You also need to add the additional variable into the future dataframe. You can use Listing 19-14 to do this.

Listing 19-14. Add reservations to the model

```
m4 = Prophet()
m4.add_regressor('reserve_visitors')
m4.fit(train4)
future4 = m4.make_future_dataframe(periods=len(test))
future4['ds'] = pd.to_datetime(future4['ds']).dt.date

future4 = future4.merge(X_reservations, left_on = 'ds', right_on =
'visit_date', how = 'left')[['ds', 'reserve_visitors']].fillna(0)
```

```
forecast4 = m4.predict(future4)

print(r2_score(list(test['y']), list(forecast4.loc[450:,'yhat'] )))

plt.plot(list(test['y']))
plt.plot(list(forecast4.loc[450:,'yhat'] ))

fig2 = m4.plot_components(forecast4)
plt.show()
```

This model obtains an R2 score of **0.849**. This is better than any of the previous configurations, and it means that the number of reservations is a good addition to the model. You can see the plot of the predictions in Figure 19-6.

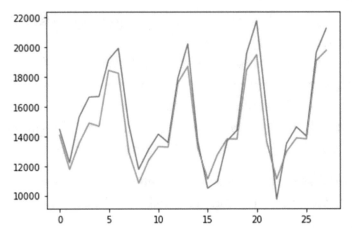

Figure 19-6. *Predictions against the actual data*

You can also see the different components fitted by this model in Figure 19-7. In this specific model, you see the impact of the additional variable "reservations."

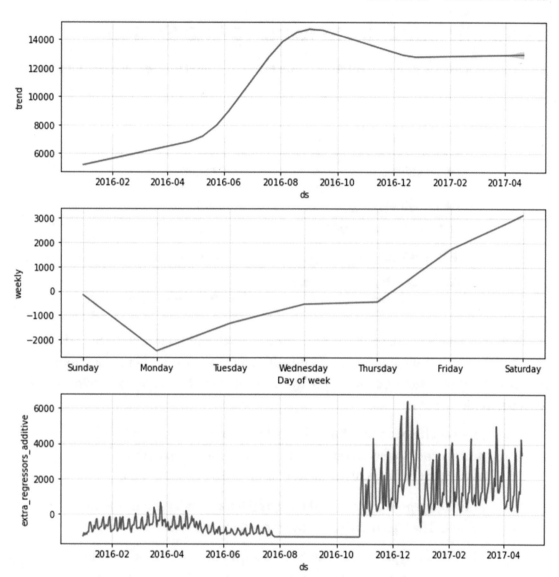

Figure 19-7. *Decomposition of the model*

Tuning Hyperparameters Using Grid Search

To recap, let's list the different variants of the Prophet model that we've seen until here in Table 19-1.

Table 19-1. *Results of the Prophet models*

Model Variant	R2 Score
Basic Prophet	**0.808**
With monthly seasonality	**0.787**
With holidays	0.806
With reservations as regressor	0.849

Now 0.849 is a great score. We have now covered all additions that we can add to the model: extra seasonality, holidays, and additional regressors. There is still room for improvement by tuning a number of hyperparameters. The hyperparameters of the Prophet model are as follows:

- The **Fourier order** can be set for each seasonality. A higher Fourier order means higher frequency of changes, which means that seasonality curves with a higher Fourier order are less smooth.

- The **changepoint_prior_scale** allows the trend to be fit more or less flexible. The higher the value, the more flexible the trend.

- The **holidays_prior_scale** allows holidays' effect to be less important. The default value is 10, and the lower it gets, the less important the holidays are.

- The **seasonalities** also have a **prior scale**. For this tuning example, we will leave them out; but know that you can tune them as well.

We will do a grid search of those hyperparameters to see whether this can improve the model. When we are at it, we might as well add the previous tests into the equation again. After all, there is a possibility that when changing some of the new hyperparameters, this changes the optimum on the other choices.

As covered before, a basic approach to grid search is to loop through any combination of values for hyperparameters and then select the best-performing combination. Listing 19-15 shows how to go about this.

Listing 19-15. Grid searching Prophet

```
def model_test(holidays, weekly_seasonality,
        yearly_seasonality, add_monthly, add_reserve, changepoint_prior_
        scale, holidays_prior_scale, month_fourier):

    m4 = Prophet(
            yearly_seasonality=yearly_seasonality,
            weekly_seasonality=weekly_seasonality,
            holidays=holidays,
            changepoint_prior_scale=changepoint_prior_scale,
            holidays_prior_scale=holidays_prior_scale)

    if add_monthly:
        m4.add_seasonality(
            name='monthly',
            period=30.5,
            fourier_order=month_fourier)

    if add_reserve:
        m4.add_regressor('reserve_visitors')

    m4.fit(train4)

    future4 = m4.make_future_dataframe(periods=len(test))

    future4['ds'] = pd.to_datetime(future4['ds']).dt.date

    if add_reserve:
        future4 = future4.merge(
            X_reservations,
            left_on = 'ds',
            right_on = 'visit_date',
            how = 'left')
```

```
        future4 = future4[['ds', 'reserve_visitors']]
        future4 = future4.fillna(0)

    forecast4 = m4.predict(future4)

    return r2_score(
            list(test['y']),
            list(forecast4.loc[450:,'yhat'] ))
# Setting the grid
holidays_opt = [holidays, None]
weekly_seas = [ 5, 10, 30, 50]
yearly_seas = [ 5, 10, 30, 50]
add_monthly = [True, False]
add_reserve = [True, False]
changepoint_prior_scale = [0.1, 0.3, 0.5]
holidays_prior_scale = [0.1, 0.3, 0.5]
month_fourier = [5, 10, 30, 50]

# Looping through the grid
grid_results = []
for h in holidays_opt:
  for w in weekly_seas:
    for ys in yearly_seas:
      for m in add_monthly:
        for r in add_reserve:
          for c in changepoint_prior_scale:
            for hp in holidays_prior_scale:
              for mf in month_fourier:
                r2=model_test(h,w,ys,m,r,c,hp,mf)
                print([w,ys,m,r,c,hp,mf,r2])
                grid_results.append([h,w,ys,m,r,c,hp,mf,r2])

# adding it all to a dataframe and extract the best model
benchmark = pd.DataFrame(grid_results)
benchmark = benchmark.sort_values(8, ascending=False)
```

```python
h, w,ys, m, r, c,hp,mf,r2 = list(benchmark.iloc[0,:])

# Fit the Prophet with those best hyperparameters
m4 = Prophet(
            yearly_seasonality=ys,
            weekly_seasonality=w,
            holidays=h,
            changepoint_prior_scale=c,
            holidays_prior_scale=hp)

if m:
    m4.add_seasonality(
            name='monthly',
            period=30.5,
            fourier_order=mf)

if r:
    m4.add_regressor('reserve_visitors')

m4.fit(train4)

future4 = m4.make_future_dataframe(periods=len(test))

future4['ds'] = pd.to_datetime(future4['ds']).dt.date

if r:
    future4 = future4.merge(
            X_reservations,
            left_on = 'ds',
            right_on = 'visit_date',
            how = 'left')
    future4 = future4[['ds', 'reserve_visitors']]
    future4 = future4.fillna(0)

forecast4 = m4.predict(future4)
```

This tuned model gives an R2 score of **0.928**. The model uses the trend, the weekly and yearly seasonality, together with the number of reservations as an extra regressor. The weekly seasonality order is **50**, and the yearly seasonality Fourier order is **10**. The "changepoint_prior_scale" is **0.1**, and the "holidays_prior_scale" does not matter as holidays aren't used in the final model. You can see the predictive performance in Figure 19-8, and the decomposition of this final model is shown in Figure 19-9.

It may be possible that a better result lies outside of the grid. Feel free to change the grid parameters and rerun the grid search to see if you can obtain an even better score.

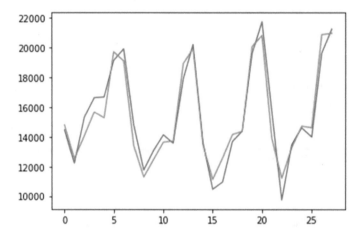

Figure 19-8. *Predictive performances of the tuned model*

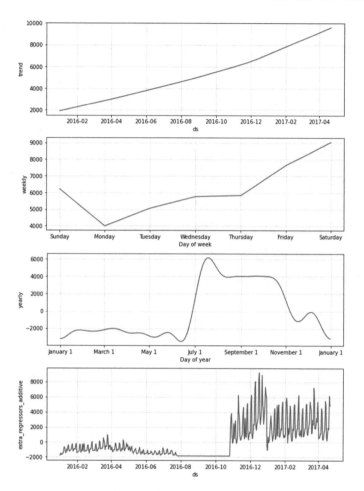

Figure 19-9. *Decomposition of the tuned model*

Key Takeaways

- Facebook Prophet is an automated procedure for building forecasting models developed by Facebook.

- The model is easy to set up and tune, which is a great advantage.

- Input possibilities are

  - Seasonality of any regular order

  - Holidays

  - Additional regressors

- You can use Prophet for univariate models only. For multivariate models, you need to build multiple Prophet models.

- The model's goal is to obtain good results without too much intervention, but there are a few hyperparameters that you can tune:

  - *The Fourier order of the seasonality*: A higher order means more flexibility.

  - *The* changepoint_prior_scale *plays on the trend*: The higher the value, the more flexible the trend.

  - *The* holidays_prior_scale: The lower it is, the less important the holidays are for the model.

  - *The prior scale for the seasonality*

CHAPTER 20

The DeepAR Model

DeepAR is a model developed by researchers at Amazon. DeepAR provides an interface to building time series models using a deep learning architecture based on RNNs. The advantage of using DeepAR is that it comes with an interface that is easier to use for model building when compared to Keras.

DeepAR can be considered a competitor with Facebook's Prophet that you have seen in the previous chapter. Both models try to deliver a simple-to-use model interface for models that are very complex under the hood. If you want to obtain good models with relatively little work, DeepAR and Prophet are definitely worth adding to your model benchmark.

About DeepAR

DeepAR forecasts univariate or multivariate time series using RNNs. The theoretical added value of DeepAR is that it fits a single model on all the time series at the same time. It is designed to benefit from correlations between multiple time series, and therefore it is a great model for multivariate forecasting. Yet it can also be applied to a single time series.

DeepAR models multiple types of seasonality variables. It also creates variables for the day of the month, day of the year, and other derived variables. As a user of the DeepAR model, we do not have much control over those variables and the inner workings of the algorithm. The model is relatively a black box aside from a few hyperparameters that we will come back to later.

© Joos Korstanje 2021
J. Korstanje, *Advanced Forecasting with Python*, https://doi.org/10.1007/978-1-4842-7150-6_20

Model Training with DeepAR

To discover the DeepAR model, we will use the dataset on restaurant visits that was used in the previous chapter. This way, we can do a benchmark of DeepAR against Facebook's Prophet. The data are accessible on www.kaggle.com/c/recruit-restaurant-visitor-forecasting. You can prepare the data by doing the sum of all restaurants using Listing 20-1.

Listing 20-1. Importing the data

```
import pandas as pd
y = pd.read_csv('air_visit_data.csv.zip')
y = y.pivot(index='visit_date', columns='air_store_id')['visitors']
y = y.fillna(0)
y = pd.DataFrame(y.sum(axis=1))

y = y.reset_index(drop=False)
y.columns = ['date', 'y']
```

As a recap, the tuned, final, Facebook Prophet model obtained an R2 Score of 0.928. Let's see whether DeepAR is able to match this performance. We will be using the DeepAR model that is presented by the gluonts library. I recommend using the following link for installation instructions and understanding the dependencies: https://ts.gluon.ai/install.html.

To use this library, we need to start by preparing the data to be in the format that this library understands. Gluonts uses a relatively unintuitive data format. It uses an object called **ListDataset**. It must contain *the target variable, a start timestamp, and a frequency*. The train data needs to contain the train data only, whereas the test data needs to contain the train and the test data.

You can build those objects using Listing 20-2. Note: Unlogically, it is necessary to specify the seasonality "H" for hourly rather than "D" for daily to obtain a reasonably accurate forecast. This is likely due to an unresolved bug in gluonts at the time of writing.

Listing 20-2. Preparing the data format required by the gluonts library

```
from gluonts.dataset.common import import ListDataset
start = pd.Timestamp("01-01-2016", freq="H")
```

```
# train dataset: cut the last window of length "prediction_length", add
"target" and "start" fields
train_ds = ListDataset([{'target': y.loc[:450,'y'], 'start': start}],
freq='H')
# test dataset: use the whole dataset, add "target" and "start" fields
test_ds = ListDataset([{'target': y['y'], 'start': start}],freq='H')
```

Now, let's start building a simple DeepAR model using gluonts in Listing 20-3. To do this, you need to create an **estimator** called the **DeepAREstimator**, which in turn uses a **trainer.** The important arguments in this function are the following:

- The **prediction_length**, which is the number of steps that you want to predict.

- The **ctx** is where you can specify whether you want to run on CPU or on GPU. Running on GPU is much faster, but you can do this only if you have a GPU available on your hardware.

- The number of **epochs** is the number of times that you want all the data to be passed through the underlying deep neural network.

- The **learning rate** is the step size for the optimizer of the neural network: a larger learning rate allows you to make larger steps, but it can miss the optimum, whereas a smaller learning rate may make the optimization slow or let you get stuck in a local optimum.

Listing 20-3. Fitting the default DeepAR model

```
from gluonts.model.deepar import DeepAREstimator
from gluonts.trainer import Trainer
import mxnet as mx
import numpy as np

np.random.seed(7)
mx.random.seed(7)

estimator = DeepAREstimator(
    prediction_length=28,
    context_length=100,
    freq='H',
    trainer=Trainer(ctx="gpu", # remove if running on windows
```

```
                    epochs=5,
                    learning_rate=1e-3,
                    num_batches_per_epoch=100
                    )
)

predictor = estimator.train(train_ds)
```

Predictions with DeepAR

Then to make predictions on the test set, you can use Listing 20-4. There is something complicated going on in this step. The model doesn't simply make one prediction, but rather many predictions. It makes, by default, a hundred predictions per time step, each an alternative future trajectory.

This may sound weird, as you generally want to have just one forecast. This forecast should be the most accurate possible. Yet the idea is having forecasted a large number of potential trajectories, you can take the median of those 100 alternative trajectories as a forecast. The median is also called the 0.5th quantile.

Listing 20-4. Prediction

```
predictions = predictor.predict(test_ds)
predictions = list(predictions)[0]
predictions = predictions.quantile(0.5)
```

Let's compute the test R2 score and make a plot to see what those predictions look like using Listing 20-5.

Listing 20-5. R2 score and prediction graph

```
from sklearn.metrics import r2_score
print(r2_score( list(test_ds)[0]['target'][-28:], predictions))

import matplotlib.pyplot as plt
plt.plot(predictions)
plt.plot(list(test_ds)[0]['target'][-28:])
plt.legend(['predictions', 'actuals'])
plt.show()
```

The R2 score that this model obtains is **0.83**. Be aware that DeepAR predictions are not the same every run, even though the random seed has been set. You may obtain a different outcome. You can see the prediction graph in Figure 20-1.

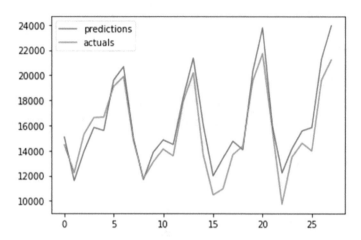

Figure 20-1. *Predictions of the default DeepAR forecast*

Probability Predictions with DeepAR

Since DeepAR builds many possible trajectories of one and the same forecast, this allows to make a similar graph, but that adds a confidence interval around the curve. You can use Listing 20-6 to obtain the graph, and it is shown in Figure 20-2.

Listing 20-6. Probability forecast graph

```
from gluonts.evaluation.backtest import make_evaluation_predictions

forecast_it, ts_it = make_evaluation_predictions(
    dataset=test_ds,   # test dataset
    predictor=predictor,   # predictor
    num_samples=100,   # number of sample paths we want for evaluation
)

forecasts = list(forecast_it)
tss = list(ts_it)

ts_entry = tss[0]
```

```
forecast_entry = forecasts[0]

def plot_prob_forecasts(ts_entry, forecast_entry):
    plot_length = 150
    prediction_intervals = (50.0, 90.0)
    legend = ["observations", "median prediction"] + [f"{k}% prediction
    interval" for k in prediction_intervals][::-1]

    fig, ax = plt.subplots(1, 1, figsize=(10, 7))
    ts_entry[-plot_length:].plot(ax=ax)   # plot the time series
    forecast_entry.plot(prediction_intervals=prediction_intervals,
    color='g')
    plt.grid(which="both")
    plt.legend(legend, loc="upper left")
    plt.show()

plot_prob_forecasts(ts_entry, forecast_entry)
plt.show()
```

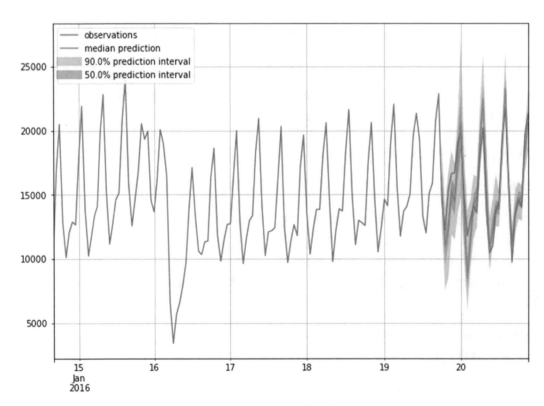

Figure 20-2. *Plot of the probabilities around the forecast*

Adding Extra Regressors to DeepAR

To go further with DeepAR, let's discover how to add additional regressors to the model. In many cases of forecasting, you have information about the future that can improve the quality of your forecast. In the case of the restaurant visits forecast, we have two additional datasets: the holidays and the restaurant reservations.

You can imagine that those will help the model to learn better. The advantage of using a model like DeepAR is that the only thing that you have to do is to update the ListDataset to include those additional variables and it will automatically model them. The model itself will decide which balance to make between seasonality and additional regressors. Let's see how to specify the ListDataset with additional regressors in Listing 20-7.

Listing 20-7. Preparing holiday and reservation data and adding them into the ListDataset

```
X_reservations = pd.read_csv('air_reserve.csv.zip')
X_reservations['visit_date'] = pd.to_datetime(X_reservations['visit_
datetime']).dt.date
X_reservations = pd.DataFrame(X_reservations.groupby('visit_date')
['reserve_visitors'].sum())
X_reservations = X_reservations.reset_index(drop = False)

# Convert to datatime for merging correctly
y.date = pd.to_datetime(y.date)
X_reservations.visit_date = pd.to_datetime(X_reservations.visit_date)

# Merging and filling missing dates with 0
y = y.merge(X_reservations, left_on = 'date', right_on =  'visit_date',
how = 'left').fillna(0)

# Preparing and merging holidays data
holidays = pd.read_csv('date_info.csv.zip')
holidays.calendar_date = pd.to_datetime(holidays.calendar_date)
y = y.merge(holidays, left_on = 'date', right_on = 'calendar_date',
how = 'left').fillna(0)

# Preparing the ListDatasets
```

```
train_ds = ListDataset([{
    'target': y.loc[:450,'y'],
    'start': start,
    'feat_dynamic_real': y.loc[:450,['reserve_visitors', 'holiday_flg']].
    values
    }], freq='H')

test_ds = ListDataset([{
    'target': y['y'],
    'start': start,
    'feat_dynamic_real': y.loc[:,['reserve_visitors', 'holiday_flg']].
    values
    }],freq='H')
```

Now to use this model, you can simply use the exact same code as before: the change has to be done only in the ListDataset as a feat_entry_real. DeepAR will automatically understand this as an additional regressor, and it'll know what to do with it. As proof, Listing 20-8 fits the model and prints the R2 score of the model with regressors holidays and reservations.

Listing 20-8. Same code for fitting a different model: this model contains the two additional regressors

```
np.random.seed(7)
mx.random.seed(7)

# Build and fit model
estimator = DeepAREstimator(
    prediction_length=28,
    context_length=100,
    freq='H',
    trainer=Trainer(ctx="gpu", # remove if running on windows
                    epochs=5,
                    learning_rate=1e-3,
                    num_batches_per_epoch=100
                    )
)
```

```
predictor = estimator.train(train_ds)

# Make Predictions
predictions = predictor.predict(test_ds)
predictions = list(predictions)[0]
predictions = predictions.quantile(0.5)

# Compute and print R2 score
print(r2_score( list(test_ds)[0]['target'][-28:], predictions))
```

The R2 score obtained by this model is **0.872**. Be aware that your result may vary due to uncontrolled randomness. This R2 score is better than the R2 score of the model without the additional regressors, so let's keep them in our dataset while moving forward.

Note   Often, when running models on CPU vs. GPU or when using distributed computation, it is very difficult to control randomness. This can be a disadvantage, but GPU and distributed computation make the execution so much faster that it is often worth it to lose the control over your randomness. After all, the randomness is, at least theoretically, not supposed to play a role in obtaining good results. When you have inconsistent results, the best thing is to test multiple runs and try to estimate how much variation you observe. If this is too much for having reliable forecasts, you can always switch to simpler models or try and tweak the model to become more stable.

Hyperparameters of the DeepAR

Now it's time to look at some hyperparameters and see how this model can be tuned. Three hyperparameters will be added into a small grid search loop to see whether they can improve the R2 score:

- learning_rate

- num_layers: The number of RNN layers (default: 2)

- num_cells: The number of RNN cells for each layer (default: 40)

You can use Listing 20-9 to do a loop through a selection of values for those three hyperparameters.

Listing 20-9. Tuning the hyperparameters

```
np.random.seed(7)
mx.random.seed(7)

results = []

for learning_rate in [1e-4, 1e-2]:
  for num_layers in [2, 5]:
    for num_cells in [30, 100]:

      estimator = DeepAREstimator(
          prediction_length=28,
          freq='H',
          trainer=Trainer(ctx="gpu", # remove if on Windows
                      epochs=10,
                      learning_rate=learning_rate,
                      num_batches_per_epoch=100
                  ),
          num_layers = num_layers,
          num_cells = num_cells,
      )

      predictor = estimator.train(train_ds)

      predictions = predictor.predict(test_ds)

      r2 = r2_score(list(predictions)[0].quantile(0.5), list(test_ds)[0]
      ['target'][-28:])
      result = [learning_rate, num_layers, num_cells, r2]
      print(result)
      results.append(result)
```

The best model coming out of this grid search has a learning rate of **0.01**, with **two** RNN layers and **100** cells per layer. This model obtains an R2 score of **0.874**.

This hyperparameter search is relatively short on purpose, as the model takes quite some time to run. As a more general rule, here is an overview of all the hyperparameters that you could try to tune according to the DeepAR literature:

- learning_rate between 1e-5 and 1e-1

- num_layers between 1 and 8 (default: 2)

- num_cells between 30 and 200 (default: 40)

- epochs between 1 and 1000

- context_length: The number of steps to unroll the RNN before computing predictions between 1 and 200 (default: none)

- dropout_rate: The dropout regularization parameter between 0 and 0.2 (default: 0.1)

Benchmark and Conclusion

Now using DeepAR, you were able to obtain an R2 score of **0.874**. This is great, but it is not beating the best-tuned version of Facebook's Prophet, which obtained **0.928**.

We've observed something difficult in DeepAR: it is hard to control randomness in this model. When making the forecast, DeepAR makes a large number of potential trajectories, and the median of those can be used as final forecast. Yet when running the code a second time, those trajectories are not the same, and therefore their median will differ a bit as well. This could be considered a disadvantage of DeepAR.

If you want to use this model for accurate forecasting, you will need to analyze the variability of the model thoroughly. DeepAR is a powerful model and can also be used as a great analysis tool. Obtaining a large number of trajectories can also give you very useful boundaries of probability. In some cases, it may be much more useful to understand the different scenarios that could occur rather than have one, most accurate forecast.

For example, imagine that you're forecasting the number of restaurant visitors to estimate whether you'll make enough money to open a second restaurant. In this case, it may be very valuable to have a lower bound of the number of visitors. If you would go bankrupt with this low scenario, you know that you take a serious risk if you'd choose to open the restaurant.

For forecasting accurately, there are still other elements to consider: DeepAR is much more suitable than Prophet for doing multivariate forecasting. This was not included in the benchmark as it is simply impossible with Prophet. With Prophet you'd have to develop and tune a model for each time series, whereas one tuned DeepAR model can work on all the time series at once.

In the next chapter, we'll do a final overview of model benchmarking and model selections to understand which data and other decision rules to use for those decisions.

Key Takeaways

- DeepAR is a model built by Amazon that builds deep forecasts based on RNNs.

- The gluonts package implements DeepAR in the DeepAREstimator.

- The DeepAR model can learn trends from many time series at the same time and make forecasts of multiple time series.

- When predicting, it predicts a large number of potential trajectories. You can use the median of those trajectories as forecast.

- DeepAR has a large component of randomness, and it is difficult to obtain reproducible results. This can be a no-go depending on the exact use case that you are working on.

CHAPTER 21

Model Selection

In the last 18 chapters, you've seen a long list of machine learning models that can be used for forecasting. In this chapter, you'll find some final reflections on how to decide which of the models to use for your practical use cases.

Model Selection Based on Metrics

The book started with an overview of model metrics. Throughout the different chapters, a number of model benchmarks have been done based on R2 scores. This was done firstly in Chapter 15, where XGBoost performances were benchmarked against LightGBM performances. Another benchmark was presented in Chapter 17, between the SimpleRNN, GRU, and LSTM.

The benchmarks presented in this book were based on metrics. When using this approach, there are two general approaches. A first approach is to build a train and a test dataset. You then tune and optimize a number of models on the train data, after which you compute the R2 score on the test data. The model that performs best on the test data is selected.

The second common way to do metrics-based model selection is to use cross-validation. When you have little data, it can be costly to keep out data for a test set, and cross-validation can be a solution to have metrics-based model selection. Another advantage of using cross-validation is that each model is trained and tested multiple times, so randomness is less influential. After all, the test could be favorable to a certain model, purely due to the random selection process and bad luck.

Yet in practice, there are other things than metrics to consider when choosing the right model for your use case. Let's see a few of them.

© Joos Korstanje 2021
J. Korstanje, *Advanced Forecasting with Python*, https://doi.org/10.1007/978-1-4842-7150-6_21

Model Structure and Inputs

Another thing to think about when selecting a model is that it needs to fit with the underlying variation of your dataset. For example, it does not make sense to use a model with autoregressive components if you know empirically that there is no autoregressive effect in your use case. Let's do a quick recap of the types of components that can be modeled in the different chapters of this book.

We started with **univariate time series models**, which allow forecasting one variable based on past variation in itself. Those models are useful when you either don't have any other data to use in the modeling process or, more importantly, when there really is **a time series component** (like autoregression, moving average, etc.) present in the data.

After that, we covered **multivariate time series models**, which do the same thing but applied to multiple time series at the same time. They have an advantage when you need to forecast multiple time series at the same time, as you have to build only one model for multiple time series rather than building a model for each. Also, it can benefit from shared variation between the time series and have an advantage in performance.

Then you've seen how to use a number of classical **supervised machine learning models** for forecasting. From a high level, those models all work in the same way. They convert a number of input variables into a target variable. To add seasonality to those models, you can convert the seasonality information into input variables. This type of model is great if you have not just a target variable, but you also have other information about the future that you can use for forecasting, for example, using the number of restaurant reservations when trying to forecast restaurant visits.

We then moved on to using multiple types of **Neural Networks**. The first type of Neural Network that we saw was a **Dense Feedforward Neural Network**. This neural network is really similar to the way that classical supervised machine learning models work: there are input variables and one (or multiple) target variable(s).

The other Neural Networks that we have seen were **Recurrent Neural Networks**. They are quite different as they are made to learn and predict sequences of data. The input data for RNNs have therefore to be prepared as sequences. You can also add external variables into RNN models.

Neural Networks can be very performant when parameterized well. Yet tuning and parameterizing a neural network can be hard: it takes a lot of time from the developer, but also computation time to get to a good result.

Lastly, you've seen two modeling techniques from **large technology companies**. You have seen Facebook's Prophet and Amazon's DeepAR. Those models are generally using neural networks under the hood. Those approaches try to automate complex technology with easy-to-use interfaces.

In some ways this works: it is easier to tune Prophet or DeepAR than tuning a neural network. In other ways, there is still work to do: we have seen that tuning is still necessary to obtain a great result. It is therefore not totally automated.

Prophet and DeepAR are also able to learn both seasonal components and additional data. Prophet is not able to model multiple time series at once, whereas that is possible with DeepAR.

One-Step Forecasts vs. Multistep Forecasts

Related to the question of models being able to use additional regressors is the question of models being able to do multistep forecasts. As we've seen, there are basically three categories of models.

The first category of models can simply not do forecasts of multiple steps. This is the case for models that need the latest data point each time that you want to predict a next data point.

The second category is when the model can do multistep forecast by iterating over the forecasted values and redoing one-step forecasts every time. In this case, it is taking the forecasted value as a past data point at the risk of adding up more and more errors.

The third category can easily learn multistep forecasts. In some cases, this is by doing one multivariate forecast in which you treat each time step as a new dependent variable. An alternative is to learn relations between input sequences and output sequences.

Model Complexity vs. Gain

Besides metrics and the types of variables that can be used by a model, I want to come back to model complexity. For some models, it takes a long time to obtain a slight improvement in R2. You should wonder if, for your use case, there is a way to say that a certain improvement of accuracy is still worth it. For example, spending a year for 0.001 points of R2 improvement is probably not worth it when doing sales forecasting, yet it may be totally worth it when doing medical work or aircraft safety or things for the army.

You must consider that model tuning costs time for a modeler, but it also costs computing time, which can become expensive when running, for example, cloud computing setups for very long periods.

This type of consideration can be seen when we moved from classical supervised models to Neural Networks. Neural Networks are very performant and can obtain great results. It is much more complex to parameterize and optimize a neural network architecture than to do a hyperparameter tuning of, for example, a gradient boosting. In short, you would need to have an idea of the impact of your R2 score to be sure how your forecast still has a practical impact on your use case.

Model Complexity vs. Interpretability

A second problem with complex models is that it often becomes very difficult to understand what the models have learned. For univariate time series or for classical supervised models, it is relatively easy to extract the model coefficients or to use other tools for model interpretation like printing a decision tree or extracting variable importance.

For more complex models, including multivariate time series, but mainly neural networks, it is much harder to go through huge coefficient matrices and understand in humanly understandable terms what it is that the model has learned.

For Prophet and DeepAR, model interpretability is even worse, as the black box approach of those models has put the model builder even further away from the actual technicalities of what has been learned.

Model interpretability is important. As a forecaster, you need to be confident in your model. To be confident, you should at least be totally sure of what it does. This is often not the case with complex models and can lead to very difficult situations.

Model Stability and Variation

As a last reflection for choosing a model for your forecasts, I want to come back to the question of model stability. In Chapter 20, we have observed that DeepAR is great at giving a confidence region in which we may expect our model to be, but that when repeating the model training, it is generally not capable of reproducing the same result. This problem of model stability may be a serious problem when you are trying to make a reliable forecast.

Model variation is also important when talking about overfitting models. We have seen that when models overfit, they can get to learn the training data by heart, and therefore they are not learning a general truth anymore, but rather they are learning noise of the training data.

This is also a type of variation: when you expect a great result based on your modeling process, but you obtain bad results when forecasting in practice, this may be a variation in metrics that is due to overfitting. In a way, it is also a model metric. It may not be the one we generally talk about first, yet it is still a very important concept to keep in mind when moving from the development phase to the actual forecasting on the field.

Conclusion

Throughout the book, you have seen how to manage model benchmarks, and you have seen different techniques for model evaluation, including the train-test set, cross-validation, and more. Until here, model performance has been generally defined by a metric, like the R2 score. Those metrics show how good a model performs on average.

The average performance is generally the first thing to consider, but in this chapter, you have seen a number of additional reflections on model selection that are also important to have during the model development phase.

Having gone through the first 20 chapters of this book, you should have all necessary input for making performant forecasts using Python. The general reflections posed in this chapter should help you to avoid some easily made yet important mistakes.

After all, when you make a forecasting model, the most important thing is not whether it works only on your training data and not even whether it works on your test data. When applying in practice, all that matters is whether your prediction for the future comes true.

Key Takeaways

- Model metrics generally tell us whether a model is good on average.

- Model variation tells us whether there is a lot of variation in a model's quality metric, for example, if we retrain it on a slightly different dataset.

- More complex models can sometimes give better model metrics, but they often come at a cost of time spent on modeling, computing costs, and a cost of lower model interpretability.

- When selecting a model type, you should take into account whether the model you use is capable of fitting the types of effects that you would theoretically expect to be present in your dataset, including the possibility for adding extra regressors and the possibility to do multistep forecasts.

Index

A

Activation function, 210, 211, 242

Aggregation, 180, 181

ARMA model, 89
 ACF and PACF plots, 93, 94
 actual *vs.* fitted values, 96
 cross-validation, 100, 101
 development processes, 89
 grid search, 100–103
 hyperparameters, 100
 hyperparameter tuning, 99
 KPI, 96–99
 mathematical definition, 90
 metrics, 95–97
 past predictions, 89
 Python source code, 90
 sunspot data, 91, 92

Artificial Intelligence (AI), 3

Augmented Dickey Fuller (ADF) test, 51, 52, 77, 92, 96, 134

Autocorrelation
 differenced data, 54–56
 Earthquake dataset
  computing correlation, 49
  correlation matrix, 50
  dataset, 46
  describe method, 47
  evaluation, 48
  nan value (missing value), 50

 profile report, 47
 shifted data, 49
 lags
  autocorrelation function, 55
  correlation coefficient, 55, 56
  definition, 55
  differenced data, 55
  model evaluation and benchmarking, 59
  partial model, 57, 58
  predictive performance, 59
  statsmodels package, 55
  time series, 58, 59
 positive/negative, 50

Autocorrelation function (ACF)
 ARIMA model, 108
 ARMA model, 93

Autoregressive (AR) model, 45
 Autocorrelation (*see* Autocorrelation)
 definition, 59
 differenced data, 53–55
 stationarity/ADF test, 51, 52
 univariate time series models, 45
 Yule-Walker method, 60–68

Autoregressive, Integrated Moving Average (ARIMA) model
 CO2 data
  ACF and PACF plots, 108, 110
  hyperparameter tuning, 110–113
  importing data, 107

© Joos Korstanje 2021
J. Korstanje, *Advanced Forecasting with Python*, https://doi.org/10.1007/978-1-4842-7150-6

Autoregressive, Integrated Moving
  Average (ARIMA) model (*cont.*)
    plotting data, 107
    definition, 106
    equation, 106
    integration, 106
    linear trend, 106
    univariate time series, 105

B

Backpropagation algorithm, 211–213
Backtesting model, 39
Bagging, 180
Bayesian optimization
    compare performances, 204
    hyperparameter tuning, 200, 201
    LightGBM model, 202, 203
    scikit-optimize package, 202
    XGBoost, 203
Bootstrap aggregation, 180

C

Cross-validation
    K-fold, 34–36
    rolling time series, 38, 39
    time series models, 36–38

D

Decision Tree model
    bike-sharing, 166
    bike-sharing users, 161
    DecisionTreeRegressor, 165
    dendrogram, 167
    explanatory values, 162
    grid search, 164, 165

hyperparameter tuning, 164
if-else decisions, 159
influential outlier value, 162
intuitive, 159
mathematical/algorithmic
    components, 160
    pruning/reducing complexity, 160
    splitting, 160
max_depth, 167
parameters, 165
training dataset, 163
DeepAR model, 273, 274

E

Exclusive Feature Bundling (EFB), 196

F

Forecasting, computing power, 3
Fully connected neural networks
    schema, 209–211

G, H, I, J

Gated Recurrent Unit (GRU), 227
    hidden layers, 240–242
    history graph, 239, 241
    model architecture, 238, 239
    technical issues, 237
Gradient-Based One-Side
    Sample (GOSS), 196
Gradient boosting (XGBoost/LightGBM)
    model, 193
    AdaBoost algorithm, 195
    Bayesian optimization (*see* Bayesian
        optimization)
    boosting process

ensemble learning, 193
histogram-based splitting, 195
iterative process, 193
partial derivatives, 194
differences, 195–197
forecasting traffic data, 197, 198
LightGBM forecasting traffic
data, 199, 200
grid search, 184, 185

K

Keras library
compilation, 220
development process, 220
history plot, 225
history plots, 223
hyperparameters, 219
model architecture, 220
source code, 223, 224
training history, 221
train-test split, 219
Key performance indicator (KPI), 96–99
K-fold cross-validation, 34–36
k-nearest neighbors (kNN) model
alternative predictions, 172, 173
grid search, 175
intuitive explanation, 169
mathematical definition, 169–171
prevalent methods, 171
random search, 176
traffic data, 172–175
weighted heavier, 171

L

LightGBM (gradient boosting
algorithm), 195
Linear regression

CO_2 dataset
drop missing values, 155
graphical data, 152
lagged variables, 155
predictive performance, 154, 156
Python source code, 151
variables, 152, 153
matrix notation, 151
model definition, 150
OLS model, 151
variables, 149
visual interpretation, 150
Long Short-Term Memory
(LSTM), 227, 243
definition, 243
one-layer network, 246–248
performances, 244–246
sigmoids/tanh operations, 243, 244
three layers/64 cells, 248–250
training history, 248

M

Machine learning techniques
classification *vs.* regression, 17
description, 4
landscape, 4
supervised model, 9–17
time series models, 4–9
univariate *vs.* multivariate
models, 18
unsupervised *vs.* supervised
models, 17
Mean Absolute Error (MAE), 27, 220
Mean Absolute Percent
Error (MAPE), 28
Mean Squared Error (MSE), 25, 26
Metrics, 24, 25

Model evaluation, 21
    backtesting model, 39
    combination, 40–42
    explanatory variables, 41
    hypothetical data, 21–24
    strategies
        cross-validation, 34–36
        errors, 29
        overfitting, 30
        train-test split, 30, 31
        validation split, 32–34
Model selection
    complexity *vs.* gain, 287
    interpretability, 288
    metrics, 285
    one-step *vs.* multistep forecasts, 287
    stability *vs.* variation, 289
    structure/inputs, 286
Moving Average (MA) model
    actuals *vs.* forecast, 79, 80
    AR model, 74, 75
    autocorrelation function, 78
    differenced data, 77
    fitting model, 73, 74
    graphical data, 76
    grid search, 85–87
    impulses, 86
    model definition, 72
    multistep forecasting, 82–84
    optimization methods, 73
    out-of-sample forecast, 81
    parameters, 74
    partial autocorrelation function, 79
    stationarity, 74
    stock price data, 76
    train data and evaluation, 81
    univariate time series, 71
    Yahoo Finance package, 76

N

Negative autocorrelation, 51
Neural Networks (NNs)
    activation layers, 211
    backpropagation, 211
    data preparation
        MinMaxScaler code, 216
        PCA model, 216–219
        scaling/standardization, 215
        scree plot, 218
    data preparation methods, 214, 215
    fully connected neural networks
        schema, 210, 211
    hyperparameter tuning, 213, 214
    keras, 219–225
    learning rates, 212
    optimizers, 212
Non-stationary, 51, 52
Normal distribution, 97

O

Ordinary Least Squares (OLS)
    method, 61, 151
Overfitting models, 30, 42

P, Q

Partial autocorrelation function (PACF),
    57, 74, 94
    ARIMA model, 109
    ARMA model, 93
Positive autocorrelation, 50
Prinicipal component analysis (PCA),
    216–219
Prophet model
    advantage/disadvantage, 253
    data format, 254, 255

decomposition plot, 257, 258

DeepAREstimator, 275

dependent variable, 254

Facebook's Prophet, 253

forecast plot, 256, 257

gluonts library, 274

grid search, 266–271

holiday data, 260–262

hyperparameters, 266, 281–283

Kaggle page, 254

ListDataset, 279, 280

prediction, 255, 256

predictions, 277, 278

probability forecast graph, 277, 278

R2 score, 255

regressors, 26–265, 279–281

reservations, 263

seasonality, 259, 260

training model, 274–276

univariate forecasts, 254

univariate/multivariate time series, 273

R

Random Forest/XGBoost model, 179

  distributions

    max_features, 186, 187

    n_estimators, 187, 188

    RandomizedSearchCV, 188, 189

    search option, 185

    uniform distribution, 187

  ensemble learning, 180, 181

    ensemble learning, 180

  feature importance, 190

  hyperparameter tuning, 184

  interpretation, 189–191

  intuitive data, 179

  sunspots, 182–184

  variable subsets, 182

Recurrent Neural Networks (RNNs), 227

  architecture, 227, 228

  data preparation, 230–233

  vs. dense network, 228

  GRU model, 237–241

  lagged variables, 229

  MinMaxScaler, 231

  multivariable, 230

  prediction, 230

  SimpleRNN (*see* SimpleRNN layer)

Rolling time series model, 38, 39

Root Mean Squared Error (RMSE), 26

R2 (R squared) metrics, 28, 29

S

SARIMAX model

  mathematical definition, 126

  time series, 125

  vs. supervised models, 126

  Walmart dataset, 127

    correlation matrix, 128

    endog and exog, 129

    graphical data, 127

    hyperparameter search, 129

    predictive performance, 130

    source code, 127

  X component, 125

Seasonal Autoregressive Integrated Moving
    Average (SARIMA) model, 115

  components, 115

  model definition, 116, 117

  Walmart sales

    fitting model, 120–122

    grid search, 121

    importing data, 117

    predictive performance, 119–121

SimpleRNN layer
    architecture, 228, 229
    hidden layers, 235–237
    parameterize network, 233–235
Stationarity, 51, 52
Stochastic Gradient Descent (SGD), 212
Supervised machine learning
    correlation coefficient, 16, 17
    explanatory variables, 9–17
    historical data, 9–15
    zooming view, 14, 15

T

Time series models
    cross-validation, 36–38
    explanatory variables, 7, 9
    differenced data, 53–55
    graphical representation, 6
    historical developments, 4
    hypothetical approach, 5
    linear data, 5
    Python source code, 6, 8
    seasonality, 7

U

Underfitting models, 42
Univariate time series, (*see* Time series
    models)
Univariate *vs.* multivariate models, 18

V, W

Vector autoregression moving average
    exogenous variables (VARMAX)
    model
  components, 141

hyperparameters, 143
mathematical model, 142
multivariate time series
    modeling, 142–144
Vector autoregression (VAR) models
    coefficients, 135
    differencing/integration, 134
    forecasting Walmart sales
      fitting model, 137
      maxlags, 138
      time series, 136
    hyperparameters, 134
    model definition, 133
    multiple univariate *vs.* multivariate
      model, 135
    multivariate model, 133
    stationarity, 134

X

XGBoost, 195

Y, Z

Yule-Walker method
    autocovariance
      function, 60
    coefficients, 60, 62
    equations, 60
    hyperparameters, 66
    Kronecker delta function, 60
    matrix format, 60, 61
    OLS method, 61
    source code, 62, 63
    stand-alone model, 68
    strategies and metrics, 63
    train-test evaluation, 64–68
    underfitting, 66

Printed in the United States
by Baker & Taylor Publisher Services